T0213693

# SpringerBriefs in Electrical and Computer Engineering

More information about this series at http://www.springer.com/series/10059

Huaqing Zhang • Sami Khairy • Lin X. Cai
Zhu Han

# Resource Allocation in Unlicensed Long Term Evolution HetNets

Huaqing Zhang
Department of Electrical and Computer
Engineering
University of Houston
Houston, TX, USA

Lin X. Cai
Department of Electrical and Computer
Engineering
Illinois Institute of Technology
Chicago, IL, USA

Sami Khairy
Department of Electrical and Computer
Engineering
Illinois Institute of Technology
Chicago, IL, USA

Zhu Han
Department of Electrical and Computer
Engineering
University of Houston
Houston, TX, USA

ISSN 2191-8112          ISSN 2191-8120   (electronic)
SpringerBriefs in Electrical and Computer Engineering
ISBN 978-3-319-68311-9          ISBN 978-3-319-68312-6   (eBook)
https://doi.org/10.1007/978-3-319-68312-6

Library of Congress Control Number: 2017956897

Printed on acid-free paper

This Springer imprint is published by Springer Nature
The registered company is Springer International Publishing AG
The registered company address is: Gewerbestrasse 11, 6330 Cham, Switzerland

# Contents

# Introduction

## 1 Long Term Evolution Over Unlicensed Band (U-LTE)

The phenomenal increase in mobile subscribers and the rich multimedia applications lead to an unprecedented demand for broadband access in next generation wireless networks. It is expected that the next generation of 5G networks will support many new killer applications, such as 4K Ultra High Definition (UHD) video streaming, virtual reality based applications, and massive Internet of Things (IoT). UHD TV requires a bandwidth on the order of tens of megabits per second, and is replacing cable and satellite TV nowadays. Virtual reality based applications are emerging as an integral part of the workplace, education, healthcare and entertainment. Moreover, billions of IoT devices such as wearable electronics, household devices and sensors, pave the way for smart homes and cities, changing the way people lives. It is not surprising that Cisco predicts that the global mobile data traffic will increase seven folds by 2021, reaching 49 exabytes per month, with 78% of which being video traffic [1].

To fulfill the unprecedented demand and provision Quality of Service (QoS) for high densities of mobile users, wireless service providers aim to develop advanced solutions to further augment network capacity at the minimum cost. It is well recognized that dense deployment of a multi-tier heterogeneous network (HetNet), including macro-cell, micro-cell, pico-cell, and femto-cell, is a desirable and feasible solution for increasing the spatial network capacity and QoS provisioning of 5G networks, by exploiting enhanced inter-cell interference coordination (eICIC) techniques. Generally, small cell technology allows energy efficient communications over a shorter distance with a lower power consumption, compared with conventional communications in macro-cells. Besides licensed small cells, Wi-Fi operating over the unlicensed band is also considered as an alternative small cell solution for traffic offloading from the licensed band to the unlicensed band in Long Term Evolution (LTE). Due to the different access technologies

© The Author(s) 2018
H. Zhang et al., *Resource Allocation in Unlicensed Long Term Evolution HetNets*,
SpringerBriefs in Electrical and Computer Engineering,
https://doi.org/10.1007/978-3-319-68312-6_1

in the unlicensed band, it is hard to coordinate the transmissions in Wi-Fi and cellular networks. Thus, Wi-Fi is usually loosely coupled in LTE HetNets to provide complementary capacity. Recently, the Federal Communication Committee (FCC) opened extra sub-bands in the 60 GHz band for unlicensed use, which opens a door for a new research activity in the 3rd Generation Partnership Project (3GPP) [2], to exploit the unlicensed spectrum bands along with the existing licensed band to provide enhanced capacity in LTE networks, based on the existing LTE network architecture [3–5]. Unlicensed LTE (U-LTE) is considered as a promising solution to provide high user performance and seamless user experience under a unified radio technology by extending LTE to the readily available unlicensed spectrum.

## 2   Research Challenges

Extending LTE to unlicensed bands is by no means a simple task, due to the different characteristics of the licensed and unlicensed bands. LTE was originally designed to operate on licensed spectrum bands, which are exclusively used by the owner operator. The main objective of LTE operators is to maximize the spectral utilization efficiency of the expensive licensed band to provision seamless mobile services with guaranteed QoS to mobile users [6]. To achieve this goal, LTE adopts centralized Radio Resource Management (RRM) in multi-tiered heterogeneous networks (HetNets), comprising different types of network cells [7–12] to manage the interference among licensed users. However, unlicensed spectrum is usually shared by various unlicensed systems of different access technologies, given that the FCC transmission regulations are met. For example, the ISM (Industrial, Scientific and Medical) band in the 2.4 GHz is shared by various devices including IEEE 802.11 Wi-Fi, IEEE 802.15.1 Bluetooth, IEEE 802.15.4 Zigbee, and cordless phones for a variety of applications, such as wireless internet access, manufacturer monitoring and automation, and telemedication. Compared with the congested 2.4 GHz band, the relatively under-utilized 5 GHz U-NII (Unlicensed National Information Infrastructure) unlicensed band is mainly used for indoor Wi-Fi networks, and also attracts the attention of various wireless internet service providers. Therefore, the foremost issue in unlicensed systems is to allow various unlicensed users to efficiently and friendly coexist with each other without causing severe interference [13, 14]. Due to the difficulty in finding a common central controller for different unlicensed systems, unlicensed users are usually distributively coordinated to access the unlicensed spectrum, and there is no QoS guarantee in the unlicensed band. In addition, the stringent FCC regulations limits the transmit power of unlicensed users so that unlicensed users can only communicate over a limited distance in a local area, while licensed LTE is not subject to such FCC regulations and can provide seamless broadband services to mobile users in a wide area. As an unlicensed system integrated in the licensed LTE, U-LTE should not only comply with unlicensed regulations and distributively coordinate with other unlicensed systems, but should also be tightly coupled with the centralized LTE to provide seamless broadband services to mobile users with guaranteed QoS.

U-LTE should be integrated into the unified LTE network architecture and use the same network access technologies of the conventional LTE for service provisioning. That is, U-LTE will use the same core network, follow the same authentication, security and management procedures, and be well synchronized with the licensed LTE for integrated services. Similar to other unlicensed networks, U-LTE is also subject to the FCC regulations and thus is more suitable for small network cells. However, RRM for U-LTE small cells faces many new great challenges that beckons for further research:

1. The first challenge is efficient spectrum sharing and harmonious co-existence of U-LTE with various unlicensed systems, especially the widely deployed Wi-Fi network and U-LTE small cells of other operators. Harmonious co-existence means no unlicensed system will dominate or starve any other unlicensed system. As an integral part of LTE, U-LTE is inherently well synchronized for transmissions, while other unlicensed users such as Wi-Fi users may adopt asynchronous channel access. How to allow synchronous U-LTE to autonomously coordinate with various unlicensed systems with no or different time synchronization to achieve adaptive, efficient and fair spectrum sharing is an interesting yet challenging issue.

2. The second challenge is distributed coordination in U-LTE. Like LTE, U-LTE is also a multi-carrier system that exploits all available unlicensed bands for high data rate services. Unlike conventional LTE that adopts centralized RRM schemes, U-LTE requires distributed RRM schemes to coordinate multiple unlicensed users to transmit over multiple unlicensed channels, based on the performance guarantee of other unlicensed systems. Due to the lack of a common central controller for different U-LTE and other unlicensed systems, it is critical to characterize different access technologies in different channels in the distributed RRM to ensure efficient resource utilization and spectrum sharing of multiple unlicensed networks.

3. The third challenge is efficient carrier aggregation of U-LTE. Generally, the existing LTE carrier aggregation (CA) technologies cannot be simply applied to aggregate unlicensed spectrum bands for centralized radio resource scheduling, because the existing CA assumes the exclusive usage of the licensed bands without considering the interference from co-existed unlicensed users. Compared with interference from the licensed users belonging to the same operator, interference from unlicensed users are more random and cannot be managed and predicted. How to appropriately manage the radio resources and offload the traffic into U-LTE HetNets to achieve better integrated services need to be investigated.

# 3 Resource Allocation in U-LTE HetNets

In this book, we systematically study radio resource allocation for U-LTE HetNets. The first research issue we target is efficient radio access management of U-LTE. We examine various coexistence schemes of U-LTE with other systems over the unlicensed band. Specifically, we study the existing coexistence technologies and develop analytical models to analyze their performance. Based on the analysis, we identify the key performance issues of these technologies with respect to fairness and protocol overheads. To address these issues, we then propose an adaptive coexistence scheme to achieve a better and more harmonious coexistence. The proposed coexistence scheme is further analyzed and the protocol parameters are fine tuned to achieve the best coexistence performance among unlicensed systems. System-level simulations are also carried out to evaluate and verify the performance of various coexistence schemes.

Spectrum sharing plays a critical role in achieving high spectrum utilization among unlicensed users. We then investigate spectrum sharing strategies in the U-LTE HetNets, considering the scenarios where multiple wireless service operators coexist simultaneously in the unlicensed spectrum. Due to the autonomous behaviors of each wireless service operator, game theory is put forward to analyze resource allocation strategies of each wireless service operator in a distributive fashion. In order to guarantee the performance of other unlicensed systems, transmit power of each wireless service operator in the unlicensed spectrum is restricted according to the behaviors of other unlicensed systems. As the transmit power of one wireless service operator can affect the utilities of other wireless service operators, competition among multiple wireless service operators is analyzed. By predicting the corresponding reactions of other operators, the optimal strategy of each wireless service operator is proposed and the Nash equilibrium solution is achieved.

As there are multiple unlicensed bands available to the U-LTE HetNets with various interference from other unlicensed systems, we further analyze the spectrum matching strategies among U-LTE HetNets and Wi-Fi systems for stable and optimal solutions. Due to various spectrum allocations in Wi-Fi system, in order to guarantee the performance of Wi-Fi users and improve the performance of all U-LTE HetNet users, the interaction between LTE and Wi-Fi users, is modeled as the stable marriage (SM) game in matching theory. Based on different preferences of U-LTE HetNet users and Wi-Fi users on different unlicensed bands, how to perform the coupling between U-LTE HetNet users and Wi-Fi users on different unlicensed band is analyzed.

In addition, from the perspective of each wireless service operator in U-LTE HetNet, how to beneficially manage the traffic offloading from the licensed bands to the unlicensed bands remains a significant issue. Due to spectrum sharing among all service operators and other unlicensed systems in the unlicensed spectrum, it is significant for each service operator to predict the behaviors of its users and the resource allocation strategies of other service operators before making decisions to optimize its own performance. Accordingly, a multi-operator multi-user Stackelberg game is proposed, where all wireless service operators act as leaders and all users act as followers. Considering both competitive and coordinative behaviors of all wireless service operators, spectrum allocation and power control strategies are achieved with optimal and equilibrium solutions.

# 4 Outline

The outline of this book is as follows. In chapter "Radio Access Management of U-LTE", we first investigate existing coexistence technologies in U-LTE, and propose a new coexistence mechanism to achieve harmonious coexistence of U-LTE and Wi-Fi. In chapter "Game Theory Based Spectrum Sharing", considering multiple wireless service operators in the unlicensed spectrum, we propose game theory based spectrum sharing strategies. For the optimal coexistence between U-LTE and Wi-Fi systems in all unlicensed spectrum, in chapter "Spectrum Matching in Unlicensed Band with User Mobility", we propose a spectrum matching for U-LTE users and Wi-Fi users with user mobility. Spectrum management and allocation for each wireless service operator when its congested data traffic is offloaded from the licensed spectrum to the unlicensed spectrum are investigated in chapter "Traffic Offloading from Licensed Band to Unlicensed Band". Finally, we conclude our book in chapter "Conclusions and Future Works" and discuss some promising future development directions.

# References

1. Cisco, "Cisco visual networking index: Global mobile data traffic forecast update, 2016–2021", Feb. 2017.
2. Q. Incorporated, "Extending LTE advanced to unlicensed spectrum," Dec. 2013.
3. R. Zhang, M. Wang, L. X. Cai, Z. Zheng, and X. Shen, "LTE-Unlicensed: the future of spectrum aggregation for cellular networks," *Wireless Communications, IEEE*, vol. 22, no. 3, pp. 150–159, Jun. 2015.
4. Qualcomm, "Making the Best Use of Unlicensed Spectrum for 1000x," Sep. 2015, White Paper.
5. J. Andrews, S. Buzzi, W. Choi, S. Hanly, A. Lozano, A. Soong, and J. Zhang, "What will 5G be?" *Selected Areas in Communications, IEEE Journal on*, vol. 32, no. 6, pp. 1065–1082, Jun. 2014.
6. R. Zhang, Z. Zheng, M. Wang, X. Shen, and L. L. Xie, "Equivalent capacity in carrier aggregation-based LTE-A systems: A probabilistic analysis," *Wireless Communications, IEEE Transactions on*, vol. 13, no. 11, pp. 6444–6460, Nov. 2014.
7. S. Bhaumik, S. Chandrabose, K. M. Jataprolu, G. Kumar, A. Muralidhar, P. Polakos, V. Srinivasan, and T. Woo, "CloudIQ: A Framework for Processing Base Stations in a Data Center," in *Proceedings of the 18th Annual International Conference on Mobile Computing and Networking*, Mobicom '12, New York, NY, USA, 2012, pp. 125–136. [Online]. Available: http://doi.acm.org/10.1145/2348543.2348561

8. J. Andrews, "Seven ways that HetNets are a cellular paradigm shift", *IEEE Communications Magazine*, vol. 51, no. 3, pp. 136–144, 2013.

9. Z. Zheng, L. X. Cai, R. Zhang, and X. S. Shen, "RNP-SA: Joint relay placement and subcarrier allocation in wireless communication networks with sustainable energy," *Wireless Communications, IEEE Transactions on*, vol. 11, no. 10, pp. 3818–3828, Oct. 2012.

10. C. S. Chen and F. Baccelli, "Self-optimization in mobile cellular networks: Power control and user association," in *IEEE International Conference on Communications (ICC)*, May 2010, pp. 1–6.

11. P. Kulkarni, W. H. Chin, and T. Farnham, "Radio resource management considerations for LTE femto cells," *SIGCOMM Comput. Commun. Rev.*, vol. 40, no. 1, pp. 26–30, Jan. 2010. [Online]. Available: http://doi.acm.org/10.1145/1672308.1672314

12. S. Abd El-atty and Z. Gharsseldien, "On performance of HetNet with coexisting small cell technology," in *Wireless and Mobile Networking Conference (WMNC), 2013 6th Joint IFIP*, Apr. 2013.

13. "LTE-U Forum," 2014, Formed by Verizon in cooperation with Alcatel-Lucent, Ericsson, Qualcomm Technologies, Inc., a subsidiary of Qualcomm Incorporated, and Samsung. [Online]. Available: http://www.lteuforum.org/.

14. LTE-U Forum, "LTE-U SDL Coexistence Specifications," Jun. 2015.

# Radio Access Management of U-LTE

## 1 Introduction

5G is expected to support new killer applications such as 4K Ultra High Definition (UHD) video streaming, virtual reality based applications, Internet of Things (IoT) and wireless sensor networks. UHD TV, which requires a bandwidth on the order of tens of megabits per second, is replacing cable and satellite TV nowadays. Virtual reality based applications may pave the way to become an integral part of the workplace, education, healthcare, engineering and architecture; transforming life as we know it. Billions of IoT devices such as wearable devices, cameras, cars, household devices, sensor networks, just to name a few, will be added over the next 5 years. That said, it is not surprising that the global mobile data traffic is expected to increase sevenfolds by 2021, reaching 49 exabytes per month, with 78% of which being video traffic [1]. To keep up with this massive demand on wireless broadband multimedia services, 5G should therefore significantly outperform 4G in terms of bandwidth, user throughput, coverage and latency.

Two main solutions have been recognized to improve the user throughput of wireless cellular networks: cell densification and opportunistic operation over the unlicensed spectrum. Dense deployment of heterogeneous networks (HetNets) improves spectral efficiency through frequency reuse, and generally may allow for more Line Of Sight (LOS) communication links with users. With LOS links, energy efficient communications can be achieved, i.e., high data rate service at a low transmission power. U-LTE, which is the main focus of this book, is another promising solution that boosts data rates through Carrier Aggregation (CA), and achieves high spectral efficiency, due to the synchronous, scheduling-based, channel access nature of LTE [2].

**LTE Frame Structure** In order to maintain synchronization and schedule channel access for different types of information and users, LTE has defined two main frame structures for its air interface, type 1, which is applicable to Frequency

© The Author(s) 2018                                                                 7
H. Zhang et al., *Resource Allocation in Unlicensed Long Term Evolution HetNets*,
SpringerBriefs in Electrical and Computer Engineering,
https://doi.org/10.1007/978-3-319-68312-6_2

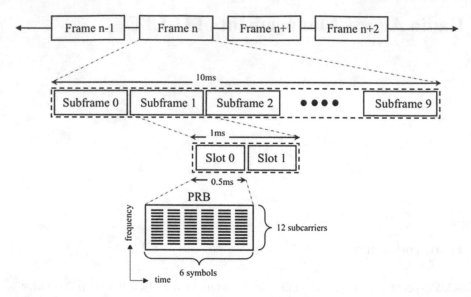

**Fig. 1** Type 1 frame structure

Division Duplex (FDD) LTE mode, and type 2, which is applicable to Time Division Duplex (TDD) LTE mode [3–5]. Type 1 frames have an overall duration of 10 ms, divided into 10 subframes of 1 ms each. Each subframe is further divided into two 0.5 ms slots, slot 1 and slot 2, as shown in Fig. 1. Each slot contains six or seven Orthogonal Frequency Division Multiplexing (OFDM) symbols,[1] and 12 subcarriers per symbol. Each 1 *subcarrier* × 1 *symbol*, is called a Resource Element (RE), and each 12 subcarriers over the duration of 1 slot form a Physical Resource Block (PRB), which is the unit for radio resource allocation in LTE. A group of PRBs with a common modulation and coding scheme (MCS) forms a transport block. The number of concurrent PRBs in a transport block depends on the bandwidth option of LTE, which may be 1.4, 3, 5, 10, 15 or 20 MHz. Therefore, multiple users can be serviced within one transport block. Such composition facilitate synchronization and resource allocation. Uplink and downlink transmissions are separated in the frequency domain in type 1 frames.

In type 2 LTE TDD mode, each 10 ms frame is divided into two half-frames of 5 ms each. Each half-frame consists of four 1 ms standard subframes, and one special subframe of 1 ms for synchronization and control information, as shown in Fig. 2. Each standard subframe is further divided into two 0.5 ms slots. The recurrence of the special subframe defines the switch-point periodicity, which can be either 5 or 10 ms. In case of 5 ms switch-point periodicity, the special subframe exists in both half-frames, whereas in the case of 10 ms switch-point periodicity, the special subframe exists in the first half-frame only. Different configurations are defined in the standard for uplink and downlink allocation of standard subframes [3].

---

[1] The number of OFDM symbols in a slot depends on the length of the cyclic prefix (CP).

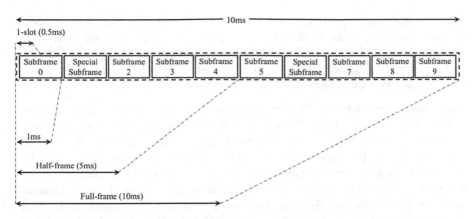

**Fig. 2** Type 2 frame structure (5 ms switch-point periodicity)

**U-LTE Modes of Operation** Depending on the availability of a licensed carrier and unlicensed supplementary links, three modes of operation for U-LTE can be identified [6]:

- **Supplemental Downlink (SDL)**. In SDL, the licensed carrier serves as the anchor carrier, carrying control plane and QoS user plane traffic. The unlicensed carrier is opportunistically aggregated when available, for high data rate downlink transmissions over a wider bandwidth. Since the control plane is reliably carried over the licensed carrier, SDL provides a secure, reliable communication with seamless mobility.
- **CA time division U-LTE**. Similar to SDL, the control plane and QoS user plane traffic are carried over the licensed spectrum. However, the unlicensed spectrum is used as a Time Division Duplex (TDD) channel when available, to carry both downlink and uplink user plane traffic.
- **Standalone U-LTE**. In standalone U-LTE, no licensed carrier is available to serve as an anchor, and both control and user plane traffic are carried over the unlicensed spectrum. Standalone U-LTE exploits the synchronous MAC of LTE and its physical layer to outperform Wi-Fi in terms of the spectral efficiency, data rates and coverage. It is worth to mention that the MulteFire Alliance is in the process of standardizing standalone U-LTE operation [7].

In all of the previous modes, LTE operates over the unlicensed band. Unlike the licensed spectrum, which is exclusively used by the spectrum owner, the unlicensed spectrum is shared by different systems with different access technologies. Since Wi-Fi is the world-wide dominant technology operating in the 2.4 and 5 GHz bands, it is imperative to design a coexistence mechanism for U-LTE, to ensure efficient, fair and harmonious operation among U-LTE and Wi-Fi systems over the unlicensed band. While LTE uses a scheduling-based Medium Access Control (MAC) protocol for synchronous transmissions, and data transmissions always start at the beginning of a superframe; Wi-Fi uses a distributed asynchronous MAC based on Carrier

Sense Multiple Access with Collision Avoidance (CSMA/CA). Therefore, a desirable coexistence mechanism should (1) comply with the synchronous superframe structure of the LTE system, (2) well adapt to asynchronous Wi-Fi transmissions, and (3) allow friendly coexistence of multiple U-LTE of different operators, which may lack access coordination in the unlicensed band.

**Coexistence Technologies** Broadly speaking, coexistence mechanisms can be classified into two categories; *Channel Splitting Mechanisms* and *Channel Sharing Mechanisms*. In cases where the unlicensed spectrum is under-utilized and there exist interference-free channels, it is intuitive that the best coexistence can be achieved by assigning those interference-free channels for full U-LTE operation. Channel splitting coexistence can be therefore realized using dynamic channel selection algorithms, which have received a considerable attention from both academia and industry. In the context of U-LTE/Wi-Fi coexistence, Qualcomm has done a pioneering work and developed a simple, yet efficient algorithm where the interference measurements are performed at the initialization stage and periodically during operation, such that U-LTE always selects the cleanest unlicensed channel for operation [8]. It is worth to mention that channel selection may suffice to achieve a harmonious coexistence, especially in places where there is little Wi-Fi activity, such as outdoor environments.

In many other cases where no clean channel can be found, channel sharing among U-LTE and Wi-Fi becomes inevitable and co-channel coexistence schemes are required. Co-channel coexistence schemes can be generally classified into two categories: non-Listen-Before-Talk (non-LBT) [9] and Listen-Before-Talk (LBT) [10] mechanisms. In the literature, it is conventional to refer to U-LTE with opportunistic SDL operating using a non-LBT MAC as LTE-U, whereas U-LTE with opportunistic SDL operating using an LBT MAC is often called LTE Licensed Assisted Access (LAA). In 3GPP classifications, non-LBT MACs are referred to as Category-1 (no LBT), while Categories 2 through 4 describe different variants of LBT-based MACs. It is worth to mention that a fair coexistence among U-LTE and Wi-Fi can be achieved as long as a coexistence mechanism is implemented, whether it is LTE-U or LAA, as shown in [11]. The main consideration when contrasting LTE-U and LAA is in fact the regulatory requirements.

In some countries such as the United States, China, South Korea and India, there are no mandatory LBT requirements for operation over the unlicensed band. A non-LBT based MAC is proposed in [12] based on the Almost Blank Subframes (ABS) feature of LTE-Advanced, in which LTE remains silent in some frames in order to yield to 802.11 users' transmissions. This coexistence mechanism is shown to increase Wi-Fi throughput, while LTE throughput decreases due to both losing time resources during blanking, and interference with Wi-Fi users during LTE transmission. Therefore, the number of blank subframes must be adapted to achieve fairness among Wi-Fi and LTE users. Qualcomm has proposed an efficient non-LBT based MAC, called the Carrier Sensing Adaptive Transmission (CSAT) [2, 8], in accordance with existing 3GPP Rel. 12 PHY and MAC standards, to facilitate early and harmonious LTE-U deployments in non-LBT markets. In CSAT, the LTE-

U node defines a duty cycle, in which it transmits during the on period and collects channel measurements during the off period. Based on these measurements, channel activity of Wi-Fi users can be inferred, and the duty cycle is adapted accordingly to ensure the impact of LTE-U on Wi-Fi performance is no worse than another Wi-Fi Access Point (AP).

Although CSAT has shown a good coexistence performance [13], the absence of LBT restricts its deployment to non-LBT markets. In most countries, LBT is a mandatory feature for unlicensed spectrum access, and therefore LBT-based coexistence schemes are recognized as the desirable global solution. 3GPP Rel. 13 has thus studied two LBT-based access technologies for LTE LAA defined by the European Telecommunications Standards Institute (ETSI): Frame Based Equipment (FBE) and Load Based Equipment (LBE) [14]. While FBE builds on top of the synchronous frame structure of LTE and is easier for implementation, it is not suitable for supporting different LTE LAA operators. LBE on the other hand, uses a busytone signal to account for the asynchronous nature of LBT, in which channel access may not necessarily be at the subframe boundary. This busytone signal reserves the channel until data transmission starts at the next subframe boundary [10]. Several works have compared the two access schemes [15–17], and studied the performance of LBE using simulations or simplified analytical models [18–20]. Nevertheless, no existing work to the best of our knowledge, has modeled the interactions between synchronous LTE LAA transmissions and asynchronous Wi-Fi transmissions, nor the reservation overhead of LBE has ever been analyzed. Thus motivated, we analytically study the performance of LBE, identify the protocol overheads, and propose a hybrid MAC design by combining the best features of FBE and LBE, to achieve the best coexistence performance of LTE LAA and Wi-Fi.

The main contributions of this chapter are fourfold:

- An analytical model is developed to analyze the performance of the LBE MAC. The reservation overhead of the protocol is derived and the throughput of LTE LAA and Wi-Fi is analyzed.
- A hybrid LBT MAC with adaptive sleep periods for LTE LAA is proposed to enable efficient and fair coexistence of U-LTE and Wi-Fi. Specifically, sleep periods are enforced within an LTE subframe, during which Wi-Fi users can transmit without contending with U-LTE. During the active period, U-LTE operates based on CSMA/CA, broadcasting a reservation signal when LBT completes.
- An analytical model is developed to study the performance of the proposed MAC, capturing the asynchronous transmissions of Wi-Fi and the synchronous transmission of U-LTE. Based on the analysis, the sleep period and the contention window size of U-LTE are optimized to achieve the best throughput fairness and the minimal reservation overhead.
- The LBE MAC and the proposed MAC are implemented in an event driven simulation platform based on NS-3. Extensive simulations are performed to validate the analysis and demonstrate the efficiency of the proposed MAC.

The rest of this chapter is organized as follows. Existing coexistence mechanisms and related work are discussed in Sect. 2. An analytical model is developed to analyze the performance of the reservation-based LBE MAC in Sect. 3. The proposed MAC and its performance analysis are presented in Sect. 4, followed by the performance evaluation in Sect. 5. Finally, our concluding remarks and future work are presented in Sect. 6.

## 2  LBT-Based Radio Access of Unlicensed LTE

In this section, we study two existing LBT-based channel access technologies specified in the ETSI standard for synchronous and asynchronous equipment, FBE and LBE, respectively [14].

### 2.1  The FBE LBT

FBE has a fixed frame structure of on and off periods, and thus it is not demand-driven. An unlicensed user will always perform a Clear Channel Assessment (CCA) for at least $20\,\mu$s at the end of the off period [14]. If the channel is idle, the user transmits in the on period for a duration equal to the Channel Occupancy Time (COT), which is the maximum time a user can transmit on a channel without performing CCA. If the channel is busy, the user abstains from transmission and repeats the CCA at the end of the off period in the next sensing subframe, as shown in Fig. 3. As per the ETSI standard, COT must be within [1 ms, 10 ms], and the minimum off period must be at least 5% of the COT. FBE falls under Category 2 (LBT without random back-off) of the 3GPP classifications.

FBE LBT-based channel access has been extensively studied in the literature, since its synchronous and periodic nature is attractive for LTE implementation. Compared to CSAT, simulation results in [2] have shown that U-LTE small cell users can achieve a higher throughput when FBE is used, while still protecting Wi-Fi networks. Reinforcement learning techniques have been used in [21] to find the optimal COT under variable Wi-Fi traffic, and a double Q-learning algorithm

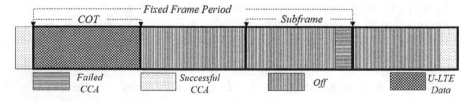

**Fig. 3**  Frame based equipment (FBE)

is proposed to enhance the throughput performance of both LTE and Wi-Fi, by jointly learning and optimizing the COT and transmit power. In [22], downlink throughput of U-LTE and Wi-Fi are mathematically analyzed using a Discrete Time Markov Chain (DTMC) model. Based on the analysis, an optimization problem is formulated to find the optimal U-LTE sensing and backoff times in presence of a given number of Wi-Fi users. An online algorithm is further proposed to adapt the access parameters. A Fair-LBT algorithm is proposed in [23] to determine the best number of idle subframes for U-LTE, considering the estimated number of Wi-Fi users, total system throughput and the fairness among users.

While FBE can achieve a friendly coexistence among U-LTE and Wi-Fi, its synchronous frame structure is not suitable for fair channel sharing among multiple synchronous, heavily-loaded unlicensed systems with no time coordination. This is because the U-LTE operator that senses the channel first is more likely to gain access to the channel, and thus starve other networks [24]. To resolve this issue, it is preferable to use random contention slots within the CCA [25], or adopt an LBE MAC.

## 2.2 The LBE LBT

LBE is a demand-driven MAC that supports channel sharing among multiple synchronous U-LTE systems, and demonstrates resilience in channel access fairness in presence of heavily loaded Wi-Fi users. In LBE, the U-LTE user first performs a CCA for at least $20\,\mu s$ whenever it has new data to transmit, and hence its asynchronous nature shows up. If the channel is idle, the U-LTE user immediately transmits a channel reservation signal, to which other contenders will backoff and abstain from accessing the channel [10, 26], as shown in Fig. 4. Data transmission starts at the beginning of the next subframe boundary, and thus the synchronization with the licensed LTE superframe is maintained, in spite of the asynchronous, demand-driven channel access requests that invoke the procedure. If the channel is busy, an extended CCA check is performed, where the channel must be observed idle for $N$ unoccupied ECCA slots for a transmission to be launched, where $N$ is a random variable uniformly generated from $[0, CW_L]$ [14]. ETSI standard defines two different variants of the extended CCA check; Option A and Option B. In Option

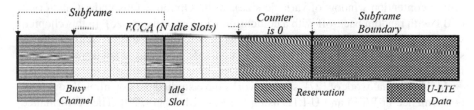

**Fig. 4** Load based equipment (LBE)

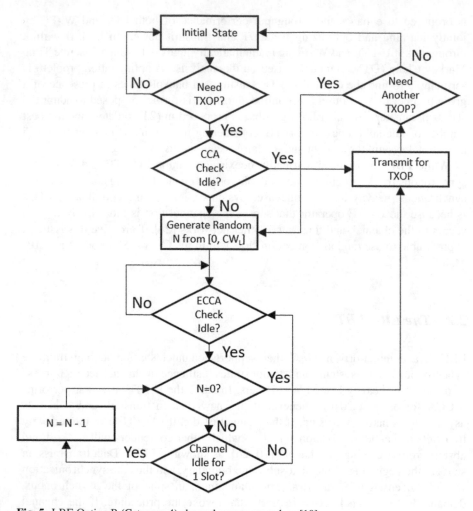

**Fig. 5** LBE Option B (Category 4) channel access procedure [10]

A, the backoff window is doubled for every failed transmission until the maximum window $CW_{max}$ is reached, whereas in Option B, the backoff window is fixed. In 3GPP classifications, Option A falls under Category 4 (LBT with random back-off with a contention window of variable size), while Option B falls under Category 3 (LBT with random back-off with a contention window of fixed size). In this chapter, we adopt Option B (Category 4) of LBE, as illustrated in the flow chart of Fig. 5, since U-LTE and Wi-Fi collisions in our model do not cause a total transmission failure of the U-LTE subframe.

In the seminal work of [27], several works have studied the downlink throughput performance of Wi-Fi and U-LTE using the LBE MAC [19, 20]. The performance of U-LTE is found to be robust in presence of Wi-Fi interference, whereas the performance of Wi-Fi drops significantly at high traffic intensities [19]. To protect Wi-Fi performance and achieve fairness, an enhanced LBT scheme based on 3GPP Category 4 was proposed in [20], adapting the contention window size of U-LTE according to channel utilization during the backoff procedure. Nevertheless, neither [19] nor [20] has captured the synchronous frame structure of LTE, and thus the reservation signal overhead is not analyzed. On the other hand, LBE with reservation based channel access is considered in the simulations of [18], where a heuristic algorithm is proposed to adapt U-LTE idle and busy durations based on channel activity statistics. The proposed algorithm is shown to improve the overall system throughput while protecting the performance of Wi-Fi under varying traffic load. Still, the overhead of channel reservation was not explicitly evaluated. Reservation-based channel access in the LBE MAC of LTE LAA can be spectrally inefficient. If U-LTE completes its LBT procedure in the beginning of a subframe for instance, a substational portion of the transmission opportunity is wasted for reservation.

Thus motivated, we first develop an analytical model to accurately character-ize the interactions between asynchronous Wi-Fi transmissions and synchronous U-LTE transmissions, quantifying the LBE protocol reservation overhead. Based on which, we propose an enhanced coexistence protocol to minimize the reservation overhead, while ensuring harmonious coexistence among U-LTE and Wi-Fi.

## 3    Performance Analysis of LBE MAC

In this section, we develop an analytical model to study the throughput performance of the LBE MAC described in Sect. 2.2.

### 3.1    System Model

A single hop WLAN with one AP and $N_w$ Wi-Fi users, coexisting with the Supplemental Downlink (SDL) of one LTE LAA network over a 20 MHz unlicensed channel, as shown in Fig. 6 is considered. In the opportunistic SDL of LTE LAA, the licensed spectrum serves as the anchor carrier and is used to transmit control plane and sensitive user plane traffic, whereas the unlicensed spectrum is opportunistically aggregated when available for additional downlink data transmission. U-LTE uses LBE MAC to access the unlicensed spectrum, whereas Wi-Fi users operate using legacy Distributed Coordination Function (DCF), based on Carrier Sensing Multiple Access with Collision Avoidance (CSMA/CA) [27, 28].

In DCF, a Wi-Fi user will sense the channel whenever it has a new frame to transmit. If the channel is sensed idle for a DIFS, the frame is transmitted, otherwise, it uniformly picks a random backoff counter from its backoff contention window. The counter is decremented every idle slot, and is frozen when the channel is sensed busy. The counter resumes when the channel is sensed idle again for a

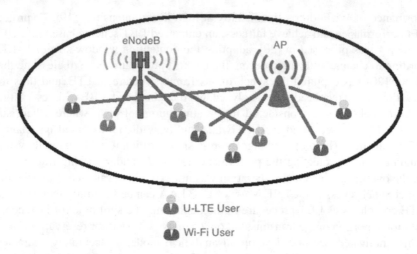

**Fig. 6** System model

DIFS. When the counter is 0, transmission starts. If the transmission is unsuccessful, the contention window is doubled, a new backoff counter is picked and channel access is started over. The backoff contention window is reset after every successful transmission, or if the maximum transmission retrials is reached, in which case the frame is dropped.

All users lie within the CCA range of each other and propagation delay is assumed to be 0. Thus, all users can sense the ongoing transmission of any other user(s) once it starts. Wi-Fi users transmit uplink data frames for a duration of $T_{PL_w}$ μs, while eNodeB's downlink data transmission takes one subframe, i.e., 1 ms, within which multiple U-LTE users are scheduled and served. All users carry saturated traffic, i.e., they always have data ready for transmission. The channel is ideal such that transmission errors only occur due to collisions between two or more simultaneous transmissions. When a collision occurs between a Wi-Fi transmission and an U-LTE transmission, the whole Wi-Fi transmission and the corresponding overlapped portion of the U-LTE transmission are corrupted.

## 3.2   Performance Analysis

Channel access in the LBE MAC can be modeled as a regenerative renewal process of idle and busy periods. Each idle and busy periods form one cycle of this renewal process, as shown in Fig. 7. During the idle period, users decrement their backoff counters synchronously, until the counter of either U-LTE or a Wi-Fi user reaches 0, in which case transmission is launched by this user on the channel. The transmission can be either a successful transmission, or a collision if multiple users transmit concurrently. Following the busy period, the channel will be idle for DIFS before users resume decrementing their counters in the next channel access cycle.

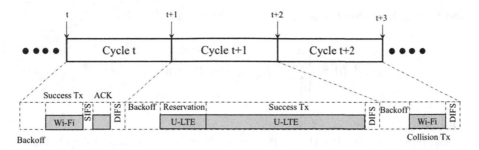

**Fig. 7** Channel access process

Let $t$ be a random time representing the beginning of any cycle, and $t + 1$ be the beginning of the next cycle, as shown in Fig. 7. The backoff processes of U-LTE and Wi-Fi users are modeled as a bi-dimensional Markov chain with states $\{s(t), b(t)\}$, representing the backoff stage and backoff counter in a cycle $t$, respectively. For the LBE MAC shown in Fig. 5, the U-LTE user has one backoff stage with a fixed window, i.e., $s_l(t) = 0$ and $b_l(t) \in [0, CW_L]$. Wi-Fi users on the other hand, operate according to CSMA/CA with binary exponential backoff (BEB); hence, the backoff states for Wi-Fi users are $s_w(t) \in [0, R]$ and $b_w(t) \in [0, CW_{s_w}]$, where $R$ is the maximum number of retransmissions and $CW_{s_w}$ is the maximum contention window at stage $s_w$. Time after $t$ is slotted into slots indexed $i = 0, 1, 2, ..$, and the duration of each slot is $\sigma$. State transitions in the Markov chain occur after each transmission, i.e., every cycle.

Let $B_l(j)$ be the steady state probability that the backoff counter of U-LTE is $j$, i.e.,

$$B_l(j) = \lim_{t \to \infty} Pr\{b_l(t) = j\}. \tag{1}$$

Define $\beta_l(i)$ as the probability that U-LTE transmits before slot $i$. Hence, $\beta_l(i)$ is the sum of the steady state probabilities of $B_l(j)$ being at any value less $i$,

$$\beta_l(i) = \begin{cases} 0, & i < 0 \\ \sum_{s=0}^{i-1} B_l(s), & 0 < i \le CW_L, \\ 1, & i > CW_L. \end{cases} \tag{2}$$

Similarly for a Wi-Fi user, $B_w(j)$ is the steady state probability that the backoff counter of a Wi-Fi user is $j$,

$$B_w(j) = \lim_{t \to \infty} Pr\{b_w(t) = j\}, \tag{3}$$

and $\beta_w(i)$ is the probability that a Wi-Fi user transmits before slot $i$,

$$\beta_w(i) = \begin{cases} 0, & i < 0 \\ \sum_{s=0}^{i-1} B_w(s), & 0 < i \leq CW_{max}, \\ 1, & i > CW_{max}. \end{cases} \tag{4}$$

where $CW_{max} = max_{s_w \in [0,R]} CW_{s_w}$.

The probability that U-LTE observes no Wi-Fi transmission before slot $i$ is thus,

$$Q_l(i) = [1 - \beta_w(i)]^{N_w}, \tag{5}$$

and the probability that a tagged Wi-Fi user observes no transmission from any other Wi-Fi user or U-LTE before slot $i$ is,

$$Q_w(i) = [1 - \beta_w(i)]^{N_w-1} * [1 - \beta_l(i)]. \tag{6}$$

Generally, $Q_l(i) - Q_l(i+1)$ is the probability U-LTE observes a Wi-Fi transmission at slot $i$, and $Q_w(i) - Q_w(i+1)$ is the probability a tagged Wi-Fi user observes a transmission by any other user at slot $i$. Without loss of generality, we first derive the state transition probabilities for the tagged Wi-Fi user, as in [29–31]. The transition from state $\{s_w(t), b_w(t)\}$ to state $\{s_w(t+1), b_w(t+1)\}$ can happen in any of the following five cases:

1. The tagged Wi-Fi user does not change its backoff stage nor decreases its backoff counter during cycle $t$, i.e., a self-loop transition in the Markov chain. This happens only if another Wi-Fi user or U-LTE, who has successfully transmitted in cycle $t - 1$, chooses a backoff counter of 0, and transmits immediately in cycle $t$. The transition probability from $\{s_w(t) = s_0, b_w(t) = b_0, b_0 \neq 0\}$ to $\{s_w(t+1) = s_0, b_w(t+1) = b_0, b_0 \neq 0\}$ is

$$P_{w1} = Q_w(0) - Q_w(1). \tag{7}$$

The transition $\{s_w(t) = s_0, b_w(t) = 0\}$ to $\{s_w(t+1) = s_0, b_w(t+1) = 0\}$ represents an impossible event, because transmission starts immediately when the backoff counter reaches 0, according to DCF rules.

2. The tagged Wi-Fi user decrements its backoff counter from $r + b_0$ to $b_0$ but does not change its backoff stage. This happens if another Wi-Fi user or U-LTE transmits in cycle $t$ after $r$ backoff slots. The transition probability from $\{s_w(t) = s_0, b_w(t) = r + b_0, b_0 \neq 0\}$ to $\{s_w(t+1) = s_0, b_w(t+1) = b_0, b_0 \neq 0\}$ is

$$P_{w2}(r) = Q_w(r) - Q_w(r+1). \tag{8}$$

The transition from $\{s_w(t) = s_0, b_w(t) = r\}$ to $\{s_w(t+1) = s_0, b_w(t+1) = 0\}$ without transmitting also represents an impossible event in DCF.

3. The tagged Wi-Fi user transmits after $r$ backoff slots, and collides with another user's transmission. Thus, the tagged Wi-Fi user goes to the next backoff stage, given $s_w(t) \neq R$. The transition probability from $\{s_w(t) = s_0(s_0 \neq R), b_w(t) = r\}$ to $\{s_w(t+1) = s_0 + 1, b_w(t+1) = b_1\}$ is

$$P_{w3}(r) = \frac{Q_w(r) - Q_w(r+1)}{1 + CW_{s_0+1}}. \tag{9}$$

4. The stage limit $R$ of the tagged Wi-Fi user is reached when it transmits and collides after $r$ backoff slots in cycle $t$. In this case, the next backoff stage is $s_w(t+1) = 0$. Thus, the transition probability from $\{s_w(t) = R, b_w(t) = r\}$ to $\{s_w(t+1) = 0, b_w(t+1) = b_1\}$ is

$$P_{w4}(r) = \frac{Q_w(r) - Q_w(r+1)}{1 + CW_0}. \tag{10}$$

5. The tagged Wi-Fi user successfully transmits after $r$ backoff slots. This happens if no other user transmits before $r+1$ slots during cycle $t$. The transition probability from $\{s_w(t) = s_0, b_w(t) = r\}$ to $\{s_w(t+1) = 0, b_w(t+1) = b_1\}$ is

$$P_{w5}(r) = \frac{Q_w(r+1)}{1 + CW_0}. \tag{11}$$

Noting that U-LTE has one backoff stage, the state transition probabilities for U-LTE can be similarly derived;

$$P_{l1} = Q_l(0) - Q_l(1),$$
$$P_{l2}(r) = Q_l(r) - Q_l(r+1),$$
$$P_{l3}(r) = P_{l4}(r) = \frac{Q_l(r) - Q_l(r+1)}{1 + CW_L}, \tag{12}$$
$$P_{l5}(r) = \frac{Q_l(r+1)}{1 + CW_L}.$$

The steady state probability for U-LTE, $\Pi_l(0,j) = \lim_{t \to \infty} Pr\{s_l(t) = 0, b_l(t) = j\}$ and for a Wi-Fi user, $\Pi_w(i,j) = \lim_{t \to \infty} Pr\{s_w(t) = i, b_w(t) = j\}$, can be then determined by numerically solving the following set of balance equations using fixed point iteration,

$$\Pi_l(0,j) = \begin{cases} \Pi_l(0,j)P_{l1} + \sum_{r=1}^{CW_L-j} \Pi_l(0,j+r)P_{l2}(r) + \\ \sum_{i=0}^{CW_L} \Pi_l(0,i)(P_{l4}(i) + P_{l5}(i)), & 0 < j \leq CW_L; \\ \\ \sum_{i=0}^{CW_L} \Pi_l(0,i)(P_{l4}(i) + P_{l5}(i)), & j = 0. \end{cases} \tag{13}$$

and,

$$
\Pi_w(s,j) = \begin{cases}
\Pi_w(s,j)P_{w1} + \sum_{r=1}^{CW_s-j} \Pi_w(s,j+r)P_{w2}(r) \\
+ \sum_{i=0}^{CW_s-1} \Pi_w(s-1,i)P_{w3}(i), \qquad 0 < s_w \le R \text{ and } 0 < j \le CW_s; \\
\\
\sum_{i=0}^{CW_s-1} \Pi_w(s-1,i)P_{w3}(i), \qquad 0 < s_w \le R \text{ and } j = 0; \\
\\
\Pi_w(0,j)P_{w1} + \sum_{r=1}^{CW_0-j} \Pi_w(0,j+r)P_{w2}(r) \\
+ \sum_{i=0}^{CW_{max}} \Pi_w(R,i)P_{w4}(i) + \\
\sum_{s=0}^{R} \sum_{i=0}^{CW_s} \Pi_w(s,i)P_{w5}(i), \qquad s_w = 0 \text{ and } 0 < j \le CW_0; \\
\\
\sum_{i=0}^{CW_{max}} \Pi_w(R,i)P_{w4}(i) + \\
\sum_{s=0}^{R} \sum_{i=0}^{CW_s} \Pi_w(s,i)P_{w5}(i), \qquad s_w = 0 \text{ and } j = 0; \\
\\
0, \qquad 0 \le s_w \le R \text{ and } CW_s < j \le CW_{max}.
\end{cases}
$$

(14)

From these steady state backoff probabilities, $B_l(j)$ and $B_w(j)$ can be found as,

$$
B_l(j) = \Pi_l(0,j) \quad \text{and} \quad B_w(j) = \sum_{s=0}^{R} \Pi_w(s,j). \tag{15}
$$

### 3.2.1  Throughput Analysis

Now that we have derived the steady state backoff probabilities of both U-LTE and Wi-Fi users, $B_l(j)$ and $B_w(j)$; respectively, we can derive the probability a cycle ends with a successful U-LTE or Wi-Fi user transmission, and the corresponding throughput performance.

The probability U-LTE transmits in a cycle, is the probability that the backoff counter of U-LTE expires *before or with* any other Wi-Fi user,

$$
P_{sl} = \sum_{i=0}^{CW_L} B_l(i)Q_l(i+1) + P_{wl}, \tag{16}
$$

where $P_{wl}$ is the probability U-LTE and a Wi-Fi user(s) transmit simultaneously at the same slot,

$$
P_{wl} = \sum_{i=0}^{CW_{max}} [Q_l(i) - Q_l(i+1)] * [\beta_l(i+1) - \beta_l(i)]. \tag{17}
$$

Given the tagged Wi-Fi user has a backoff counter $B_w(j)$, the probability it transmits successfully in a cycle is $B_w(j)Q_w(j+1)$, where $Q_w(j+1)$ is the probability all other users have a backoff counter greater than $j$ by at least one, which guarantees a successful transmission for the tagged Wi-Fi user. Hence, the probability a cycle ends with a successful Wi-Fi transmission, $P_{sw}$, is

$$P_{sw} = N_w \sum_{i=0}^{CW_{max}} B_w(i)Q_w(i+1). \tag{18}$$

From renewal theory, the normalized throughput of a user is the fraction of the average cycle duration in which the user transmits successfully. The normalized throughput of U-LTE, $S_l$, can be thus found as

$$S_l = \frac{(P_{sl} - P_{wl})T_l + P_{wl} * min(T_l, T_l + E[T_R] - T_{PL_w})}{E[cycle_1]}, \tag{19}$$

where $T_l$ is the duration of U-LTE transmission, $(P_{sl} - P_{wl})$ is the probability U-LTE transmission is collision-free, $E[T_R]$ is the expected duration of the reservation signal, $min(T_l, T_l + E[T_R] - T_{PL_w})$ is net duration in which U-LTE transmits successfully when it collides with a Wi-Fi user with probability $P_{wl}$,[2] and $E[cycle_1]$ is the expected duration of a cycle. $E[T_R]$ and $E[cycle_1]$ will be derived shortly.

The normalized reservation overhead is the fraction of the average cycle duration in which the reservation signal is transmitted,

$$S_R = \frac{P_{sl}E[T_R]}{E[cycle_1]}. \tag{20}$$

Last but not least, the normalized throughput of Wi-Fi users, $S_w$, is

$$S_w = \frac{P_{sw}T_w}{E[cycle_1]}, \tag{21}$$

where $T_w$ is the duration of a Wi-Fi transmission. In the basic access mode, a successful Wi-Fi transmission and a Wi-Fi collision have the same duration, i.e., $T_w = T_{PL_w} + SIFS + T_{ACK} = T_{PL_w} + T_{ACK_{timeout}}$. $T_{PL_w}$ and $T_{ACK}$ are the transmission time of Wi-Fi data and ACK frames, including all headers and PHY preambles; $T_{ACK_{timeout}}$ is the ACK timeout.

The expected duration of the reservation signal overheard, $E[T_R]$, is derived next. Let us consider an U-LTE transmission that ends at a subframe boundary. At the end of the backoff period, a Wi-Fi user will transmit with probability $(1 - P_{sl})$ or U-LTE will transmit with probability $P_{sl}$. If U-LTE transmits in this cycle, $T_R =$

---

[2] In case of a collision between a Wi-Fi user(s) and U-LTE, the duration of the transmission over the channel is $(T_l + E[T_R])$ since $T_l \geq T_w$, of which $min(T_l, T_l + E[T_R] - T_{PL_w})$ is collision-free.

$T_{sf} - (DIFS + E[X]\sigma)$ on average, where $T_{sf}$ is the duration of an LTE subframe. If a Wi-Fi user transmits, the next channel access cycle starts after $2*DIFS + E[X]\sigma + T_w$ on average. If U-LTE transmits then, $T_R = T_{sf} - (2 * DIFS + 2 * E[X]\sigma + T_w)$. Following this intuition, $E[T_R]$ can be compactly given by,

$$E[T_R] = \sum_{k=0}^{\infty} P_{sl}(1 - P_{sl})^k T_R(k), \tag{22}$$

where

$$T_R(k) = T_{sf} - [(k + 1)(DIFS + E[X]\sigma) + kT_w] \bmod T_{sf}, \tag{23}$$

is the expected reservation signal duration when there are $k$ consecutive Wi-Fi transmissions between any two U-LTE transmissions.

Finally, we derive the expected duration of a cycle, $E[cycle_1]$. Denote $X$ as a random variable that represents the number of initial backoff slots in a cycle. For $k$ idle slots, there is a transmission that occurs at the $k$-th slot by either U-LTE or a Wi-Fi user. The Probability Mass Function (P.M.F.) of $X$ is therefore

$$P(X = k) = \begin{cases} Q_l(k)(1 - \beta_l(k)) - Q_l(k + 1)(1 - \beta_l(k + 1)), \\ \qquad\qquad\qquad for\ 0 \leq k \leq CW_{max}, \end{cases} \tag{24}$$

and the expected number of backoff slots is

$$E[X] = \sum_{k=1}^{CW_{max}} k * P(X = k). \tag{25}$$

The expected cycle duration is therefore the sum of the expected idle period, the expected busy period, which is a weighted average of Wi-Fi and U-LTE transmissions duration, and a DIFS,

$$E[cycle_1] = E[X]\sigma + (1 - P_{sl})T_w + P_{sl}(T_l + E[T_R]) + DIFS. \tag{26}$$

## 4    Enhanced MAC Design and Performance Analysis

### 4.1    A Hybrid MAC with Adaptive Sleep

Achieving throughput fairness among multiple Wi-Fi and U-LTE operators by solely adapting $CW_L$ of the LBE MAC, does not guarantee maximum system throughput due to the reservation overhead of the protocol, which is quantified by (20). Therefore, it is desirable to design an enhanced MAC, which not only can achieve throughput fairness, but can also maximize the total system throughput.

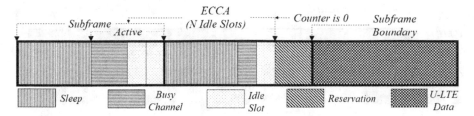

**Fig. 8**  Proposed hybrid LBT MAC

Given there are $k$ consecutive Wi-Fi transmissions between any two U-LTE transmissions, the expected duration of the reservation signal overhead $T_R(k)$, is upper bounded by the LTE subframe duration $T_{sf}$ according to (23). Thus, to minimize the protocol overhead and strike a performance balance between U-LTE and Wi-Fi, it is quintessential to configure and optimize a sleep period within each LTE subframe. To achieve this, we propose a hybrid MAC protocol, combining the best features of FBE and LBE. In each subframe of our proposed MAC, U-LTE is allowed to access the channel only in the trailing portion of a subframe, i.e., in the active period, denoted by $T_{active}$, and refrains from channel sensing in the leading portion, i.e., in the sleep period, denoted by $T_{sleep}$, as shown in Fig. 8. In our hybrid MAC, the sleep period is analogous to the mandatory idle period of the FBE MAC, while the channel access procedure during the active period is analogous to that of the LBE MAC.

Before U-LTE can transmit in the active period, it performs an initial CCA for a DIFS. If the channel is idle, U-LTE transmits a busytone signal to reserve the channel until data transmission is started at the next subframe boundary. If the channel is busy, U-LTE uniformly chooses a random number from $[0, CW_L]$. During the active period, the counter is decremented by 1 every time the channel is sensed idle for a slot of duration $\sigma$. During the sleep period, or whenever the channel is sensed busy during the active period, the counter is frozen. The counter is resumed only when the channel is sensed idle again for a DIFS during the active period. When the counter reaches 0, a reservation signal is transmitted till the subframe boundary, when an U-LTE data subframe transmission can be initiated.

This procedure is repeated whenever U-LTE has data ready for transmission. Hence, our hybrid MAC is also demand-driven like the LBE MAC. It is asynchronous in the sense that the reservation signal transmission can start at any time within the active period, yet it is synchronous in the sense that data transmission can only start at subframe boundary. Unlike the LBE MAC, reservation signal overhead is upper bounded by the active period duration, and Wi-Fi users can freely transmit without contending with U-LTE during the sleep period. By adaptively adjusting both $T_{sleep}$ and $CW_L$, the proposed MAC protocol can: (1) achieve throughput fairness between U-LTE and Wi-Fi users, (2) minimize reservation signal overhead, and (3) support the coexistence of multiple U-LTE operators, thanks to the different sleep periods and the random backoff adopted by each U-LTE operator.

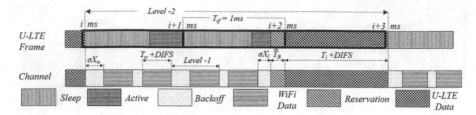

**Fig. 9** A two level renewal process model

## 4.2 Performance Analysis

The key parameters of our hybrid MAC; namely, the duration of the sleep period and the contention window size play a crucial role in achieving the best performance. To this end, we first develop an analytical model to analyze the performance of the proposed MAC; and based on the developed model, the key parameters are fine tuned to achieve the best coexistence performance in terms of throughput and fairness.

In each subframe, as shown in Fig. 9, U-LTE performs channel sensing in the active period, and freezes the backoff in the sleep period. When the backoff counter reaches 0 during the active period, U-LTE transmits a channel reservation signal, followed by one data subframe transmission. Hence, we model successive U-LTE transmissions as a two-level renewal process, where consecutive Wi-Fi transmissions are the level-1 renewal process. That is, a level-2 renewal cycle includes a random number of consecutive level-1 Wi-Fi transmissions, denoted as $N_T$, plus exactly one U-LTE transmission. Furthermore, each level-1 Wi-Fi cycle consists of: (1) a random backoff of Wi-Fi users, denoted as $X_w$, (2) one Wi-Fi Transmission $T_w$, which can be either a successful transmission or a collision, and (3) a DIFS. On the other hand, the U-LTE transmission consists of: (1) a random U-LTE backoff $X_l$, (2) a random reservation signal overhead $\hat{T}_R$, (3) the data transmission $T_l$, and (4) a DIFS.

Thus, the expected level-2 cycle duration in the hybrid MAC is

$$E[cycle_2] = E[N_T](E[X_w]\sigma + T_w + DIFS) + (E[X_l]\sigma + E[\hat{T}_R] + T_l + DIFS). \tag{27}$$

The normalized throughput U-LTE achieves when it operates using the hybrid MAC, $\hat{S}_l$, is the fraction of the level-2 cycle in which U-LTE transmits collision-free,

$$\hat{S}_l = \frac{(1 - \hat{P}_{wl})T_l + \hat{P}_{wl} * min(T_l, T_l + E[\hat{T}_R] - T_{PL_w})}{E[cycle_2]}, \tag{28}$$

where $\hat{P}_{wl}$ is the probability of a collision between a Wi-Fi user(s) and U-LTE when U-LTE transmits. The normalized reservation signal overhead $\hat{S}_R$ is,

$$\hat{S}_R = \frac{E[\hat{T}_R]}{E[cycle_2]}, \tag{29}$$

where $E[\hat{T}_R]$ is expected reservation signal duration in the hybrid MAC. Last but not least, the normalized throughput of Wi-Fi users $\hat{S}_w$, is the fraction of the level-2 cycle in which Wi-Fi users transmit successfully,

$$\hat{S}_w = \frac{E[N_T]T_w\hat{P}_s}{E[cycle_2]}, \tag{30}$$

where $E[N_T]$ is the expected number of consecutive level-1 Wi-Fi transmissions in any level-2 cycle, and $\hat{P}_s$ is the probability a Wi-Fi transmission in a level-1 cycle is successful. The successful Wi-Fi transmission probability $\hat{P}_s$ and the expected number of Wi-Fi backoff slots $E[X_w]$ can be derived as in (18) and (25) in Sect. 3.2, except that there is no U-LTE transmission, i.e., $\beta_l(i) = 0$, $\forall i \geq 0$.

To obtain $\hat{S}_l$, $\hat{S}_R$, and $\hat{S}_w$, we need to derive $\hat{P}_{wl}$, $E[X_l]$, $E[N_T]$ and $E[\hat{T}_R]$. Denote $L$ as the random backoff counter U-LTE uniformly selects from $[0, CW_L]$. The P.M.F. of $L$ is therefore given by,

$$P(L = l) = \begin{cases} \frac{1}{1+CW_L}, & 0 \leq l \leq CW_L, \\ 0, & otherwise. \end{cases} \tag{31}$$

Given U-LTE is in the active period, the probability U-LTE transmits is the probability U-LTE selects a backoff counter smaller than or equal to that of any Wi-Fi user, i.e.,

$$P(L \leq X_w) = \sum_{l=0}^{CW_L} P(L \leq X_w | L = l)P(L = l)$$
$$= \sum_{l=0}^{CW_L} \sum_{k=l}^{CW_{max}} \frac{P(X_w = k)}{1 + CW_L}, \tag{32}$$

where $P(X_w = k)$ is the probability there are $k$ idle slots in a level-1 cycle, given by (24) after setting $\beta_l(i) = 0$, $\forall i \geq 0$. Notice that $X_w$ represents the minimum of the backoff counters of all Wi-Fi users. Similarly, the probability of a collision between a Wi-Fi user(s) and U-LTE when U-LTE transmits, is the probability U-LTE selects a backoff counter equal to that of any Wi-Fi user,

$$\hat{P}_{wl} = \sum_{l=0}^{CW_L} \frac{P(X_w = l)}{1 + CW_L} \tag{33}$$

The expected number of U-LTE backoff slots $E[X_l]$ is the expected number of $L$, given U-LTE selects a counter smaller than or equal to $X_w$,

$$E[X_l] = E[L|L \leq X_w] = \sum_{k=0}^{CW_L} kP(L = k|L \leq X_w)$$

$$= \frac{\sum_{k=0}^{CW_L} kP(L = k)P(L \leq X_w|L = k)}{P(L \leq X_w)} \tag{34}$$

$$= \frac{\sum_{k=0}^{CW_L} \sum_{j=k}^{CW_{max}} kP(X_w = j)}{(1 + CW_L)P(L \leq X_w)}.$$

Channel access of U-LTE is modeled as a geometric random variable with probability of success $P(L \leq X_w)P_{active}$, where $P_{active}$ is the probability U-LTE is in the active period,

$$P_{active} = \frac{T_{sf} - T_{sleep}}{T_{sf}} = \frac{T_{active}}{T_{sf}} \tag{35}$$

Note that $T_{active} + T_{sleep} = T_{sf}$. The expected number of consecutive Wi-Fi transmissions in any level-2 cycle is therefore the number of attempts in which U-LTE fails to transmit,

$$E[N_T] = \frac{1}{P(L \leq X_w)P_{active}} - 1, \tag{36}$$

as the last transmission is from U-LTE.

Finally, the average reservation signal overhead in the hybrid MAC $E[\hat{T}_R]$, is found in a way similar to (22); however, notice that during the sleep period, the transmissions are only contributed by Wi-Fi users. Thus,

$$E[\hat{T}_R] = \sum_{k=0}^{\infty} P(L \leq X_w)(1 - P(L \leq X_w))^{h(k)}\hat{T}_R(k)I(k), \tag{37}$$

where $\hat{T}_R(k)$ is the expected reservation signal duration in the hybrid MAC when there are $k$ consecutive Wi-Fi transmissions in the level-2 cycle,

$$\hat{T}_R(k) = T_{sf} - [DIFS + k(E[X_w]\sigma + T_w + DIFS) + E[X_L]\sigma)] \bmod T_{sf}, \tag{38}$$

$I(k)$ is an indicator function that is 1 if the expected U-LTE backoff and the expected reservation signal duration of U-LTE after $k$ consecutive Wi-Fi transmissions, fall within the active period, and is 0 otherwise,

$$I(k) = \begin{cases} 1, & (E[X_L]\sigma + \hat{T}_R(k)) \leq T_{active}, \\ 0, & otherwise. \end{cases} \tag{39}$$

and $h(k)$ is the number of contentions U-LTE loses to Wi-Fi during the active periods in the level-2 cycle,

$$h(k) = \sum_{j=0}^{k} I(j) - 1. \tag{40}$$

## 4.3  Coexistence Performance Optimization

Based on the developed analytical model, we then can fine tune the duration of the sleep period $T_{sleep}$ and the contention window size $CW_L$ in the proposed MAC, to achieve the best coexistence performance between U-LTE and Wi-Fi. We design an objective function $G$ to jointly consider the total system throughput and fairness between U-LTE and Wi-Fi users,

$$G = \alpha J_F(\hat{S}_w, \hat{S}_l) + (1 - \alpha)\hat{S}_T, \tag{41}$$

where $J_F(\hat{S}_w, \hat{S}_l)$ is Jain's fairness index that evaluates the throughput fairness of U-LTE and Wi-Fi users defined by,

$$J_F(\hat{S}_w, \hat{S}_l) = \frac{(\hat{S}_w + \hat{S}_l)^2}{2(\hat{S}_w^2 + \hat{S}_l^2)}, \tag{42}$$

and $\hat{S}_T$ is the total normalized throughput of the system,

$$\hat{S}_T = \hat{S}_w + \hat{S}_l. \tag{43}$$

The objective function $G$ in (41) can be maximized by joint selection of $T_{sleep}$ and $CW_L$. $\alpha$ is a weighting factor to strike a balance between fairness and the total normalized throughput of the system. A larger value of $\alpha$ favors good fairness while a smaller $\alpha$ favors high system throughput. The optimization problem in (41) is an integer optimization problem; nevertheless, the optimal value $(T_{sleep}^*, CW_L^*)$ can be quickly found using a grid search.

## 5  Performance Evaluation

We implemented the existing LBE and our hybrid MAC in an event driven network simulator (NS-3), based on the LTE/Wi-Fi coexistence model of [32], for model validation and performance evaluation. We setup a single hop WLAN with one AP and $N_w$ saturated Wi-Fi users, coexisting with one U-LTE Base Station (eNodeB), as shown in Fig. 6. Wi-Fi users contend for access to the shared wireless channel

**Table 1** Simulation
parameters

| Parameter | Value |
|---|---|
| $T_{PL_w}$ | 244 μs |
| $\sigma$ | 9 μs |
| a DIFS | 34 μs |
| a SIFS | 16 μs |
| $T_{ACK}$ | 44 μs |
| $T_{ACK_{timeout}}$ | 44 μs |
| $CW_{min}$ ($CW_0$) | 15 |
| $CW_{max}$ | 255 |
| R | 7 |
| $T_{sf}$ | 1 ms |
| $T_l$ | 1 ms |

**Fig. 10** Normalized
throughput of LBE
($CW_L = 15$)

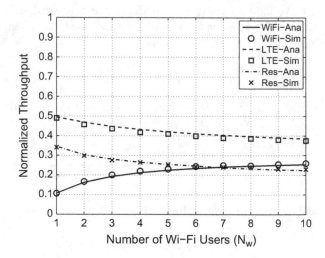

according to DCF, and transmit uplink traffic to the AP. U-LTE BS on the other
hand, contends for access using either the LBE MAC or our proposed hybrid MAC,
and whenever it wins the contention, it schedules downlink transmissions to its
associated users. The main simulation parameters are listed in Table 1.

We first study the performance of the reservation based LBE MAC with a variable
number of Wi-Fi users $N_w$ in Fig. 10. In this experiment, $CW_L = 15$. While the
normalized throughput of Wi-Fi users slowly increases as $N_w$ increases, it soon
hits a saturation wall and remains much smaller than the throughput of U-LTE.
Comparing the reservation signal overhead with the throughput of Wi-Fi users, it
can be observed that the reservation overhead is higher when $N_w \leq 6$, and is not
much less when $N_w \geq 7$. In both cases, the reservation overhead accounts for more
than 20% of the channel airtime, which demonstrates the inefficiency of the LBE
MAC.

In Fig. 11, we study the throughput performance of our hybrid MAC for fixed
values of $T_{sleep} = 700$ μs and $CW_L = 15$. With $T_{sleep} = 700$ μs, U-LTE starts

**Fig. 11** Normalized throughput of hybrid MAC with $T_{sleep} = 700\,\mu s$

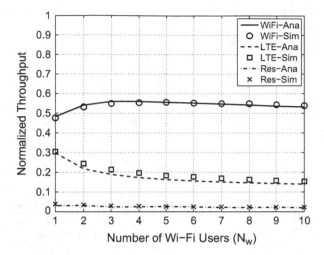

**Fig. 12** Normalized throughput of LBE ($N_w = 3$)

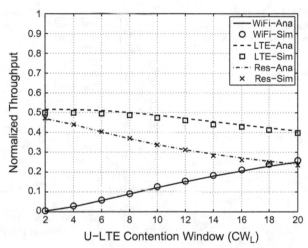

sensing the channel $300\,\mu s$ before each subframe boundary, during which it operates based on CSMA/CA. It uniformly selects a backoff counter from $[0, 15]$ and can only start broadcasting a reservation signal when its backoff counter reaches 0. It can be seen that the reservation overhead has been diminished; nevertheless, U-LTE throughput is now much lower than that of Wi-Fi due to the long sleep period. This insinuates the need to properly adjust $T_{sleep}$ in order to achieve throughput fairness between U-LTE and Wi-Fi.

In Figs. 12 and 13, we study the effects of $CW_L$ on the throughput performance of the LBE MAC and our proposed hybrid MAC; respectively, when there are $N_w = 3$ Wi-Fi users. For the LBE MAC in Fig. 12, it can be observed that the reservation signal overhead is significant, and is much higher than the throughput of Wi-Fi users, again demonstrating the inefficiency of the LBE MAC. Setting $T_{sleep} = 700\,\mu s$

**Fig. 13** Normalized
throughput of hybrid MAC
with $T_{sleep} = 700\,\mu s$
($N_w = 3$)

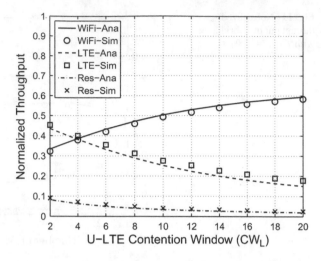

**Fig. 14** Normalized
throughput of hybrid MAC
with $CW_L = 15$ ($N_w = 3$)

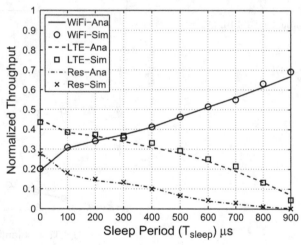

for the hybrid MAC, the reservation overhead is significantly reduced in Fig. 13. Figure 13 also shows that good fairness between U-LTE and Wi-Fi users can be achieved when $CW_L$ is set to 4.

In Fig. 14, we examine the throughput performance of 3 Wi-Fi users and U-LTE operating using the hybrid MAC with variable $T_{sleep}$ and fixed $CW_L = 15$. It is evident from Fig. 14 that the longer the sleep period in U-LTE subframes, the smaller the reservation signal overhead and U-LTE throughput are. Although U-LTE and Wi-Fi can achieve equal throughput at $T_{sleep} = 250\,\mu s$; the reservation overhead is still non-negligible and accounts for 12% of the channel airtime. Hence, it is desirable to adapt both $T_{sleep}$ and $CW_L$, in order to achieve throughput fairness at the minimal reservation overhead. Simulation results validate the accuracy of the analysis in all the figures.

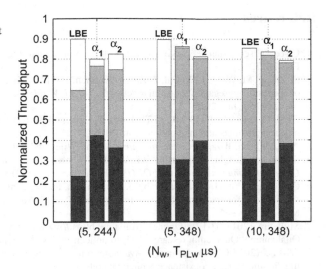

**Fig. 15** Normalized throughput of hybrid MAC at optimal $(T_{sleep}^*, CW_L^*)$ $(\alpha_1 = 0.2, \alpha_2 = 0.8)$

As can be seen from Fig. 15, our hybrid MAC can efficiently minimize the reservation overhead and maximize the total system throughput, when both $T_{sleep}$ and $CW_L$ are adapted to maximize the objective function $G$ in (41). In Fig. 15, the throughput of Wi-Fi users, U-LTE and the reservation overhead are shown under different number of Wi-Fi users $N_w$, Wi-Fi data transmission time $T_{PL_w}$, and weight factor $\alpha$. $\alpha_1 = 0.2$ favors a higher total throughput when optimizing (41), while $\alpha_2 = 0.8$ favors a greater fairness between U-LTE and Wi-Fi. The baseline for comparison is the LBE MAC with $CW_L = 15$. In all the cases, it can be seen that the total system throughput is improved and the reservation overhead is minimized.

# 6 Summary

In this chapter, we have proposed a hybrid LBT MAC for U-LTE coexisting with Wi-Fi, in order to improve the efficiency of the existing LBE MAC. We have developed an analytical framework to analyze the throughput performance of both the existing LBE MAC and the proposed hybrid-LBT MAC, considering the asynchronous Wi-Fi transmissions and the synchronous frame structure of U-LTE. The analytical model is then used to adapt the key parameters of our proposed MAC, which are the sleep period and the contention window size, such that fair coexistence between Wi-Fi and U-LTE can be achieved, and the reservation overhead can be minimized.

In our future work, we will consider multi-carrier U-LTE, where multiple non-overlapping channels in the unlicensed band are opportunistically aggregated for data transmissions over a much wider bandwidth. We plan to extend our analytical model to study the coexistence performance of multi-carrier U-LTE with Wi-Fi, with the ultimate aim of designing an efficient and fair multi-carrier U-LTE MAC.

# References

1. Cisco, "Cisco visual networking index: Global mobile data traffic forecast update, 2016–2021," Feb. 2017.
2. R. Zhang, M. Wang, L. X. Cai, Z. Zheng, X. Shen, and L. L. Xie, "LTE-Unlicensed: the future of spectrum aggregation for cellular networks," *IEEE Wireless Communications*, vol. 22, no. 3, pp. 150–159, Jun. 2015.
3. ETSI TS 136 211 (V13.0.0), "LTE Evolved Universal Terrestrial Radio Access (E-UTRA): physical channels and modulation," Jan. 2016.
4. A. F. Molisch, *Wireless Communications*. John Wiley and Sons, 2014.
5. S. Yun and L. Qiu, "Supporting Wi-Fi and LTE co-existence," in *2015 IEEE Conference on Computer Communications (INFOCOM), Hong Kong*, Apr. 2015, pp. 810–818.
6. Q. Chen, G. Yu, H. M. Elmaghraby, J. Hamalainen, and Z. Ding, "Embedding LTE-U within Wi-Fi bands for spectrum efficiency improvement," *IEEE Network*, vol. 31, no. 2, pp. 72–79, Mar. 2017.
7. Multefire. (2017-07-12). [Online]. Available: https://www.multefire.org/
8. Qualcomm. "Qualcomm research: LTE in unlicensed spectrum, harmonious coexistence with Wi-Fi". (2017-07-12). [Online]. Available: https://www.qualcomm.com/media/documents/files/lte-unlicensed-coexistence-whitepaper.pdf
9. Alcatel-Lucent, Ericsson, Qualcomm, Samsung, and Verizon, "LTE-U technical report coexistence study for LTE-U SDL V1.0," *LTE-U Forum*, Feb. 2015.
10. 3GPP TR 36.889 (V13.0.0), "Study on licensed-assisted access to unlicensed spectrum," Jun. 2015.
11. L. Simić, A. M. Voicu, P. Mähönen, M. Petrova, and J. P. D. Vries, "LTE in unlicensed bands is neither friend nor foe to Wi-Fi," *IEEE Access*, vol. 4, pp. 6416–6426, Sep. 2016.
12. E. Almeida, A. M. Cavalcante, R. C. D. Paiva, F. S. Chaves, F. M. Abinader, R. D. Vieira, S. Choudhury, E. Tuomaala, and K. Doppler, "Enabling LTE/Wi-Fi coexistence by LTE blank subframe allocation," in *2013 IEEE International Conference on Communications (ICC), Budapest, Hungary*, Jun. 2013, pp. 5083–5088.
13. A. K. Sadek, T. Kadous, K. Tang, H. Lee, and M. Fan, "Extending LTE to unlicensed band - merit and coexistence," in *2015 IEEE International Conference on Communication Workshop (ICCW), London, UK*, Jun. 2015, pp. 2344–2349.
14. ETSI EN 301 893 (V1.8.1), "Broadband radio access networks (BRAN); 5 GHz high performance RLAN," Mar. 2015.
15. Y. Li, F. Baccelli, J. G. Andrews, T. D. Novlan, and J. C. Zhang, "Modeling and analyzing the coexistence of Wi-Fi and LTE in unlicensed spectrum," *IEEE Transactions on Wireless Communications*, vol. 15, no. 9, pp. 6310–6326, Sep. 2016.
16. A. M. Voicu, L. Simić, and M. Petrova, "Inter-technology coexistence in a spectrum commons: A case study of Wi-Fi and LTE in the 5GHz unlicensed band," *IEEE Journal on Selected Areas in Communications*, vol. 34, no. 11, pp. 3062–3077, Nov. 2016.
17. C. Cano and D. J. Leith, "Unlicensed LTE/Wi-Fi coexistence: Is LBT inherently fairer than CSAT?" in *2016 IEEE International Conference on Communications (ICC), Kuala Lumpur, Malaysia*, May 2016, pp. 1–6.
18. C. K. Kim, C. S. Yang, and C. G. Kang, "Adaptive listen-before-talk (LBT) scheme for LTE and Wi-Fi systems coexisting in unlicensed band," in *2016 13th IEEE Annual Consumer Communications Networking Conference (CCNC), Las Vegas, USA*, Jan. 2016, pp. 589–594.
19. C. Chen, R. Ratasuk, and A. Ghosh, "Downlink performance analysis of LTE and Wi-Fi coexistence in unlicensed bands with a simple listen-before-talk scheme," in *2015 IEEE 81st Vehicular Technology Conference (VTC Spring), Glasgow, Scotland*, May 2015.
20. F. Hao, C. Yongyu, H. Li, J. Zhang, and W. Quan, "Contention window size adaptation algorithm for LAA-LTE in unlicensed band," in *2016 International Symposium on Wireless Communication Systems (ISWCS), Poznan, Poland*, Sep. 2016, pp. 476–480.

21. A. Galanopoulos, F. Foukalas, and T. A. Tsiftsis, "Efficient coexistence of LTE with Wi-Fi in the licensed and unlicensed spectrum aggregation," *IEEE Transactions on Cognitive Communications and Networking*, vol. 2, no. 2, pp. 129–140, Jun. 2016.
22. T. Yang, C. Guo, S. Zhao, Q. Zhang, and Z. Feng, "Channel occupancy cognition based adaptive channel access and back-off scheme for LTE system on unlicensed band," in *2016 IEEE Wireless Communications and Networking Conference, Doha, Qatar*, Apr. 2016.
23. H. Ko, J. Lee, and S. Pack, "A fair listen-before-talk algorithm for coexistence of LTE-U and WLAN," *IEEE Transactions on Vehicular Technology*, vol. 65, no. 12, pp. 10 116–10 120, Dec. 2016.
24. H. Hu, M. Zheng, K. Yu, and B. Zhou, "Enhanced listen-before-talk scheme for LTE in unlicensed band," in *2016 6th International Conference on Electronics Information and Emergency Communication (ICEIEC), Beijing, China*, Jun. 2016, pp. 168–173.
25. R. Zhang, M. Wang, L. X. Cai, X. Shen, L. L. Xie, and Y. Cheng, "Modeling and analysis of MAC protocol for LTE-U co-existing with Wi-Fi," in *2015 IEEE Global Communications Conference (GLOBECOM), San Diego, CA, USA*, Dec. 2015, pp. 1–6.
26. H. J. Kwon, J. Jeon, A. Bhorkar, Q. Ye, H. Harada, Y. Jiang, L. Liu, S. Nagata, B. L. Ng, T. Novlan, J. Oh, and W. Yi, "Licensed-assisted access to unlicensed spectrum in LTE release 13," *IEEE Communications Magazine*, vol. 55, no. 2, pp. 201–207, Feb. 2017.
27. G. Bianchi, "Performance analysis of the IEEE 802.11 distributed coordination function," *IEEE Journal on Selected Areas in Communications*, vol. 18, no. 3, pp. 535–547, Mar. 2000.
28. X. Ling, Y. Cheng, J. W. Mark, and X. Shen, "A renewal theory based analytical model for the contention access period of IEEE 802.15.4 MAC," *IEEE Transactions on Wireless Communications*, vol. 7, no. 6, pp. 2340–2349, Jun. 2008.
29. I. Tinnirello and G. Bianchi, "Rethinking the IEEE 802.11e EDCA performance modeling methodology," *IEEE/ACM transactions on networking*, vol. 18, no. 2, pp. 540–553, Apr. 2010.
30. S. Khairy, "Modeling, analysis and design of multi-channel bonding for IEEE 802.11 WLANs," Dec. 2016. [Online]. Available: http://mypages.iit.edu/~skhairy/skhairy.html.
31. S. Khairy, M. Han, L. X. Cai, Y. Cheng, and Z. Han, "Enabling efficient multi-channel bonding for IEEE 802.11ac WLANs," in *2017 IEEE International Conference on Communications (ICC), Paris, France*, May 2017, pp. 1–6.
32. L. Giupponi, T. Henderson, B. Bojovic, and M. Miozzo, "Simulating LTE and Wi-Fi coexistence in unlicensed spectrum with NS-3," *arXiv preprint arXiv:1604.06826*, 2016.

# Game Theory Based Spectrum Sharing

## 1 Introduction

As depicted in the Visual Networking Index (VNI) by Cisco [1], 5.5 billion mobile users are expected by 2021, with an average mobile connection speed of 20.4 Mbps. Compared with the 4.9 billion mobile users and 6.8 Mbps speed from 2016, the increasing number of mobile users and the threefold growth on speed motivate the exploration and expansion of other possible spectrum resources, including the unlicensed spectrum bands which are dominantly presently used by Wi-Fi networks.

Accordingly, the coexistence of Cellular Networks (CNs) and Wi-Fi networks is expected, provided that the mutual interference between CNs and Wi-Fi networks is properly under control. To address the above issue, many existing works have proposed solutions and algorithms to ensure possible coexistence of U-LTE and Wi-Fi in the unlicensed spectrum. In [2], the authors introduce the spectrum sharing problems when cellular network operators are allowed to access the unlicensed spectrum. The authors propose a hybrid method where cellular base stations can simultaneously offload traffic to Wi-Fi networks and occupy certain number of time slots on unlicensed bands. Practical strategies have been proposed to maximize the minimum average per-user throughput of each small cell. In [3], the authors introduce a network architecture where small cells can share the unlicensed spectrum with the performance guarantee of Wi-Fi systems. An almost blank subframe (ABS) scheme is employed to mitigate the co-channel interference from small cells to Wi-Fi systems, and an interference avoidance scheme is proposed based on small cell estimation of the density of nearby Wi-Fi access points. The authors in [4] evaluate and compare several existing licensed and unlicensed user coexisting mechanisms. The appropriate coexistence mechanisms, such as static muting and sensing-based adaptive, are required to achieve a balance between the performance of LTE and WLAN systems. In [5], the authors propose a cap-limited water-filling method for the U-LTE users to regulate the interference to Wi-Fi users

H. Zhang et al., *Resource Allocation in Unlicensed Long Term Evolution HetNets*,
SpringerBriefs in Electrical and Computer Engineering,
https://doi.org/10.1007/978-3-319-68312-6_3

in the unlicensed spectrum. In [6], the authors propose a novel proportional fair allocation scheme which guarantees fairness when both U-LTE and Wi-Fi coexist in the unlicensed spectrum. In [7], the authors propose a spectrum etiquette protocol to restrict the priority of U-LTE and balance the unfair competition between LTE and Wi-Fi in the unlicensed spectrum. In [8], the authors propose an "intelligent" power allocation strategy to optimize the utility of users with U-LTE and the social welfare simultaneously. In [9], an improved power control method is proposed for uplink transmissions, and thus both Wi-Fi and LTE are able to coexist with acceptable interference levels. Moreover, in order to guarantee the performance of Wi-Fi users, the strategies in cognitive radio networks can also be applied in the relations between U-LTE and Wi-Fi. In [10], the authors model the cognitive users' network access behavior as a two-dimensional Markov decision process and propose a modified value iteration algorithm to find the best strategy profiles for cognitive users. In [11], the authors jointly consider the spectrum sensing and access problems as an evolutionary game, where each secondary user senses and accesses the primary channel with the probabilities learned from its history. In [12], a Dynamic Chinese Restaurant Game is proposed to learn the uncertainties of networks and make optimal strategies. In [13], the authors propose a dynamic spectrum access protocol for the secondary users to deal with unknown behaviors of primary users. In [14], the authors investigate resource allocation problems for the uplink transmission of a spectrum-sharing-enabled femtocell network. A Stackelberg game with one leader and multiple followers is applied where the macrocell base station, i.e., the leader, sets prices to the femtocell users, i.e., the follower, to control its interference on the macrocell users. As the macrocell users and femtocell users share the licensed spectrum, each femtocell user determines and optimizes the transmit power on each sub-band only. In [15], the authors propose a fair and Quality-of-Service (QoS) based unlicensed spectrum splitting strategy to realize the joint operation of femtocell networks and Wi-Fi networks in the unlicensed spectrum band. In [16], an analytical model is developed for evaluating the baseline performance of the coexistence of Wi-Fi networks and LTE networks. In [17], a practical algorithm, which takes into account the real-time channel, interference and traffic conditions of licensed and unlicensed bands, is proposed for the integrated femto-WiFi and the dual-band femtocell to balance their traffic in both spectrum bands.

Moreover, the presence of multiple operators in a common unlicensed spectrum band makes the coexistence problem more challenging. Spectrum sharing among multiple operators has been studied in many works. In [18], the potential network efficiency gain from spectrum sharing between operators is investigated. In [19], the authors look into the problem of inter-operator sharing of radio resources, including capacity, spectrum and base stations sharing. From their work, the realistic sharing architecture and process are supported in the testbed network. However, in the unlicensed spectrum, how to jointly operate multiple wireless cellular networks and Wi-Fi networks remains a critical technical problem. Not only should we consider the competitions among all operators, but each operator is also required to ensure the performance of its users and Wi-Fi networks users at the same time. In [20], two general ideas are put forward to solve the problem. One is applying the orthogonal/exclusive use of the unlicensed spectrum for each operator. The other is to propose dynamic schemes for shared use of unlicensed radio resources. The use of unlicensed spectrum depends on the instantaneous/semi-static traffic load of U-LTE. However, the first solution lacks flexibility and the second solution requires perfect central control mechanisms.

Different from the above mentioned literature, we consider in this chapter the power control problem in a multi-operator spectrum-sharing scenario. Considering the distributive behaviors of the Wi-Fi Access Point (WAP) and each operator, game theory is introduced and applied in this scenario, so as to provide optimal strategies for each operator and Wi-Fi, to achieve high revenues. We model the interactions among all the operators and the WAP as a layered game. We first propose the zero-determinant power control strategy for a considered operator during the interaction with the WAP, by fixing the behaviors of all the other operators. With the predicted strategies of other operators, all operators play a non-cooperative game and determine their optimal power control strategies to achieve the Nash equilibrium results. Simulation results verify the theoretical analysis carried out in this chapter and show that a high performance can be achieved by applying the proposed zero-determinant strategies.

The rest of this chapter is organized in the following way. Game theory is preliminarily introduced first in Sect. 2. Then we model the system and formulate the power control problem in Sect. 3. Based on the formulated problem, we analyze the interactions between one operator and one WAP by fixing the behaviors of all other operators in Sect. 4.1. Then according to the predicted strategies between each operator and the WAP, we consider a non-cooperative game among all operators in Sect. 4.2. We present our simulation results in Sect. 5, and finally summarize our works in Sect. 6.

## 2 Preliminaries of Game Theory

Game theory is introduced as a powerful tool to analyze the distributive strategies in competitive or coordinative scenarios, which have been widely applied in economics, politics, psychology, biology, computer science, engineering, etc. With tremendous contributions, eleven game-theorists have won economics Nobel Prizes and have applied a wide range of behavioral relations among humans, animals and computers efficiently and beneficially. In game theory, there are three main characteristics, i.e., player, action and utility.

- **Player**: Players indicate the set of rational individuals which can make decisions autonomously. In the game, the conflicts normally exist among players and each player is required to make proper behaviors to either compete or coordinate with other players.

- **Action**: Actions denote the behaviors and strategies of each player during its interaction with other players. Due to conflicts, the action of one player will affect the optimal actions of other players.
- **Utility**: Utilities refer to the revenues or penalties the action brings to each player. Based on the actions of other players, each player is required to set up the optimal actions in order to achieve maximum utility for itself. Moreover, in the distributive network, as the action of other players is related to the action of the player itself, each player is required to predict and consider the possible reactions of other players, as well as determine its optimal actions to maximize its utility.

With the definition of player, action and utility, a game can be played either statically or sequentially. In the static game, all players play the game simultaneously. Accordingly, each player is required to analyze the optimal strategies of other players before determining the strategy for itself. In order to achieve stable results for all players, the Nash equilibrium concept is put forward.

**Definition 1** Let $(\mathbf{X}, \mathbf{u})$ denote the static game with $m$ players. $\mathbf{X} = \mathbf{X}_1 \times \mathbf{X}_2 \times, \ldots, \times \mathbf{X}_m$ refers to all sets of strategy profiles of all players. $\mathbf{u} = (u_1(\mathbf{x}), \ldots, u_m(\mathbf{x}))$ is the utility profile of all players. Let $\mathbf{x}_i$ be a strategy profile of player i, $\mathbf{x}_{-i}$ be a strategy profile of other players except for player i. A set of strategy profiles $\mathbf{x}^* \in X$ is able to achieve the Nash equilibrium if $\forall i$, $\mathbf{x}_i \in \mathbf{X}_i$,

$$u_i(\mathbf{x}_i^*, \mathbf{x}_{-i}^*) \geq u_i(\mathbf{x}_i, \mathbf{x}_{-i}^*). \tag{1}$$

Apart from the static game, a game can also be played sequentially. In the sequential game, the players can be divided into leaders and followers, where the leaders act first and the followers behaves correspondingly. Accordingly, the first-mover advantage exists, where the leader is able to predict the corresponding reactions of followers and make actions firstly for high utilities. In the sequential game, the stable results can be achieved with Stackelberg equilibrium, which is defined as follows.

**Definition 2** Let $((\mathbf{X}, \mathbf{A}), (g, f))$ be the general sequential game with $m$ leaders and $n$ followers. $\mathbf{X} = \mathbf{X}_1 \times \mathbf{X}_2 \times, \ldots, \times \mathbf{X}_m$ and $\mathbf{A} = \mathbf{A}_1 \times \mathbf{A}_2 \times, \ldots, \times \mathbf{A}_n$ are all sets of strategy profiles of all leaders and all followers, respectively. $g = (g_1(\mathbf{x}), \ldots, g_m(\mathbf{x}))$ is the payoff function of leaders for $\mathbf{x} \in \mathbf{X}$, and $f = (f_1(\boldsymbol{\alpha}), \ldots, f_n(\boldsymbol{\alpha}))$ is the payoff function of followers for $\boldsymbol{\alpha} \in \mathbf{A}$. Let $\mathbf{x}_i$ be a strategy profile of leader i, $\mathbf{x}_{-i}$ be a strategy profile of all leaders except for leader i, $\boldsymbol{\alpha}_j$ be a strategy profile of follower j, and $\boldsymbol{\alpha}_{-j}$ be a strategy profile of all other followers except for leader j. A set of strategy profile $\mathbf{x}^* \in X$ and $\boldsymbol{\alpha}^* \in \mathbf{A}$ is the equilibrium of the multi-leader multi-follower game if $\forall i$, $\forall j$, $\mathbf{x}_i \in \mathbf{X}_i$, $\boldsymbol{\alpha}_j \in \mathbf{A}_j$,

$$g_i(\mathbf{x}_i^*, \mathbf{x}_{-i}^*, \boldsymbol{\alpha}^*) \geq g_i(\mathbf{x}_i, \mathbf{x}_{-i}^*, \boldsymbol{\alpha}^*) \geq g_i(\mathbf{x}_i, \mathbf{x}_{-i}, \boldsymbol{\alpha}^*),$$

$$f_j(\mathbf{x}, \boldsymbol{\alpha}_j^*, \boldsymbol{\alpha}_{-j}^*) \geq f_j(\mathbf{x}, \boldsymbol{\alpha}_j, \boldsymbol{\alpha}_{-j}^*).$$

In the following sections, as all operators and the WAP are autonomous indi-viduals, which try to optimize their own utilities based on the behaviors of others, we consider all operators and the WAP as the players in one game [21]. With the established system model and formulated problems for each player, we analyze the optimal strategies of each operator and WAP pair, and the optimal strategies among all operators, respectively, in a game-theoretical perspective.

## 3  System Model and Problem Formulation

We consider an indoor environment where there is a set $\mathcal{N} = \{1, \ldots, N\}$ of operators trying to serve their MUs in the unlicensed spectrum. However, as shown in Fig. 1, the WAP already serves Wi-Fi users (WUs) in the unlicensed spectrum, so all $N$ operators are required to guarantee the performance of the WUs while increasing the QoS for their MUs. We suppose the WAP adopts Frequency Division Multiple Access (FDMA) and there are totally $S$ sub-bands in the unlicensed spectrum, each labeled as $s \in \mathcal{S} = \{1, \ldots, S\}$. As each sub-band $s \in \mathcal{S}$ of the unlicensed spectrum is independent of other sub-bands. Thus, in the following sections, we analyze the strategies of the WAP and all operators in one sub-band, say $s$, and hence drop the sub-band index to simplify notational expressions. The strategies in other sub-bands can be analyzed in a similar way. Accordingly, when the WAP shares the sub-band with all $N$ operators, the spectrum efficiency of the WAP can be expressed as

$$R^{(\mathrm{w})} = \log_2 \left( 1 + \frac{p^{(\mathrm{w})} g^{(\mathrm{w})}}{\sum_{n \in \mathcal{N}} p_n^{(\mathrm{m})} h_n^{(\mathrm{m})} + \sigma^2} \right), \tag{2}$$

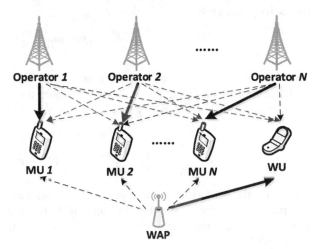

**Fig. 1** System architecture when multiple wireless operators implement LTE unlicensed in the same spectrum band (MU: mobile user; WU: Wi-Fi user)

where $p^{(w)}$ is the transmit power allocated by the WAP for a scheduled WU in the sub-band. $g^{(w)}$ is the path gain from the WAP to the WU. $p_n^{(m)}$ is the transmit power allocated by one base station (BS) of operator $n$ for a scheduled MU. $h_n^{(m)}$ is the path gain from the BS of operator $n$ to the WU. Thus, $p^{(w)}g^{(w)}$ is the signal strength that the WU receives from the WAP, and $\sum_{n \in \mathcal{N}} p_n^{(m)} h_n^{(w)}$ is the total interference from BSs of all operators. $\sigma$ is the power of the additive white noise in the sub-band.

Correspondingly, we assume that each operator serves one MU with the closest BS in the sub-band. Without causing any confusion, we shall thus interchangeably use an operator and a BS in the following analysis. The spectrum efficiency of each operator $n \in \mathcal{N}$ in the sub-band can be expressed as

$$R_n^{(m)} = \log_2 \left( 1 + \frac{p_n^{(m)} g_n^{(m)}}{p^{(w)} h_n^{(w)} + \sum_{n' \in \mathcal{N} \setminus \{n\}} p_{n'}^{(m)} h_{n'n}^{(m)} + \sigma^2} \right), \tag{3}$$

where $g_n^{(m)}$ is the path gain from the BS of operator $n$ to the scheduled MU of operator $n$. $h_n^{(w)}$ is the path gain from the WAP to the MU. $h_{n'n}^{(m)}$ is the path gain from an operator $n' \in \mathcal{N} \setminus \{n\}$ to the MU. Accordingly, $p_n^{(m)} g_n^{(m)}$ is the signal strength the MU gets from its associated BS of operator $n$, $p^{(w)} h_n^{(w)}$ is the interference the MU receives from the WAP, and $\sum_{n' \in \mathcal{N} \setminus \{n\}} p_{n'}^{(m)} h_{n'n}^{(m)}$ is the interference the MU receives from other operators in the sub-band.

Furthermore, the data transmissions from the WAP and all $N$ operators to their WU and MUs consume transmit power. To encourage minimizing power consumption, we suppose the transmit power cost of the WAP is

$$c^{(w)} = p^{(w)} r^{(w)}, \tag{4}$$

where $r^{(w)}$ is the price of unit transmit power of the WAP. The cost of transmit power for each operator $n \in \mathcal{N}$ is

$$c_n^{(m)} = p_n^{(m)} r_n^{(m)}, \tag{5}$$

where $r_n^{(m)}$ is the price of unit transmit power of the base station of operator $n$.

Therefore, in line with the above discussions, the utility of the WAP can be denoted as the achieved capacity by serving the WU minus its transmit power cost, i.e.,

$$U^{(w)} \left( p^{(w)} \big| \mathbf{p}^{(m)} \right) = BR^{(w)} - c^{(w)}, \tag{6}$$

where $B$ is the bandwidth of the considered sub-band $s$ of the unlicensed spectrum. The transmit powers of both the WAP and all WOs can affect the final utility of the WAP due to spectrum sharing. We choose $\mathbf{p}^{(m)}$ to denote the transmit powers from all operators.

Similarly, the utility of an operator $n \in \mathcal{N}$ is the achieved capacity by serving the MU minus the corresponding transmit power cost, that is,

$$U_n^{(m)} \left( p_n^{(m)} \middle| p^{(w)}, \mathbf{p}_{-n}^{(m)} \right) = BR_n^{(m)} - c_n^{(m)}, \tag{7}$$

where $\mathbf{p}_{-n}^{(m)}$ denotes the transmit powers from all operators except for the operator $n$.

We suppose that the WAP and all operators are autonomous individuals. In order to achieve a high utility for itself, the WAP should determine its transmit power $p^{(w)}$ based on the transmit powers from all operators in the sub-band. For each operator, however, based on the behaviors of all other operators and the WAP, it is supposed to determine the transmit power $p_n^{(m)}$ in order to improve its utility while guaranteeing the performance of the WU at the same time. As the WAP and all operator are able to make decisions in an iterated way, for simplicity of the analysis, we suppose the WAP and all operators have two power level choices, namely, $p^{(w)} \in \{p_1^{(w)}, p_2^{(w)}\}$, $p_n^{(m)} \in \{p_1^{(m)}, p_2^{(m)}\}$, where 1 stands for the low power level and 2 refers to the high power level. Accordingly, in the current iteration, if the probability of $p^{(w)} = p_i^{(w)}$ is $v_i^{(w)}$, and the probability of $p_n^{(m)} = p_{j_n}^{(m)}$ is $v_{j_n}^{(m),n}$, $\forall i, j_n \in \{1, 2\}$, $\forall n \in \mathcal{N}$, the expected utility of the WAP and the operator $n$ can be, respectively, shown as

$$E^{(w)} = \sum_{i, \{j_n | n \in \mathcal{N}\}} \left[ v_i^{(w)} \prod_{n \in \mathcal{N}} v_{j_n}^{(m),n} U^{(w)} \left( p_i^{(w)} \middle| \left( p_{j_n}^{(m)} \middle| n \in \mathcal{N} \right) \right) \right], \tag{8}$$

and

$$E_n^{(m)} = \sum_{i, \{j_n | n \in \mathcal{N}\}} \left[ v_i^{(w)} \prod_{n \in \mathcal{N}} v_{j_n}^{(m),n} U_n^{(m)} \left( p_{j_n}^{(m)} \middle| p_i^{(w)}, \left( p_{j_{n'}}^{(m)} \middle| n' \in \mathcal{N} \setminus \{n\} \right) \right) \right]. \tag{9}$$

For each pair of an operator $n \in \mathcal{N}$ and the WAP, if in the current iteration the transmit power of the operator $n$ is in level $x_n$, and the transmit power of the WAP is in level $y$, we define the expected probability that in the next iteration the operator $n$ decides the power in level $x_n'$ is $z_{yx_nx_n'}$, $\forall y, x_n, x_n' \in \{1, 2\}$. Correspondingly, we define the probability that in the next iteration the WAP transmits in level $y'$ is $a_{yx_1x_2...x_Ny'}$, $\forall y, x_n, y' \in \{1, 2\}$, $\forall n \in \mathcal{N}$, given that in the current iteration the transmit power of each operator $n$ is in level $x_n$, and the transmit power of the WAP is in level $y$. Accordingly, in the current iteration, the strategy profile for the operator $n$ can be given by $\mathbf{z}_n = \{z_{yx_nx_n'}, \forall y, x_n, x_n' \in \{1, 2\}\}$, $\forall n \in \mathcal{N}$. The strategy profile for the WAP can be denoted as $\mathbf{a} = \{a_{yx_1x_2...x_Ny'}, \forall y, x_n, y' \in \{1, 2\}, \forall n \in \mathcal{N}\}$.

In the iterated scenario, for an operator $n \in \mathcal{N}$, to guarantee the performance of the WU, it is required to maximize the total utility accumulated over both itself and the WAP in the same sub-band of the unlicensed spectrum, without knowing the strategy of the WAP. Furthermore, to achieve a high utility performance for the MUs, the utility of operator $n$ should be $k$ times larger than the utility of the WAP, where $k > 0$ is a constant. Eventually, the optimization problem for operator $n$ can be formulated as follows,

$$\max_{\mathbf{z}_n} \; E_n^{(\mathrm{m})} + E^{(\mathrm{w})}$$

$$\text{s.t.} \begin{cases} \mathbf{0} \leq \mathbf{z}_n \leq \mathbf{1}, \\ E_n^{(\mathrm{m})} \geq kE^{(\mathrm{w})}. \end{cases} \tag{10}$$

Based on the formulated problem, in the following sections, game-theoretical analysis is adopted to determine the optimal strategies for each operator or WAP so as to achieve its optimal utility, respectively.

# 4 Game Analysis

In this section, we analyze the optimal power control strategies for each operator and the WAP. As the strategy of an operator is affected by all other operators, we first fix the behaviors of all the other operators and discuss the optimal strategies for one operator and WAP pair in Sect. 4.1. Furthermore, by predicting the behaviors of every operator and WAP pair, each operator $n$, $\forall n \in \mathcal{N}$, is able to adjust its strategy and compete with other operators. Accordingly, in Sect. 4.2, we formulate the competition among all operators as a non-cooperative game, and find out the Nash equilibrium of the game where each of the operators cannot unilaterally change its behaviors for a higher utility.

## 4.1 Game Analysis Between an Operator and WAP

In order to better analyze the relationship between an operator $n$ and the WAP, we fix the transmit powers of all other operators, i.e., $\mathbf{p}_{-n}^{(\mathrm{m})}$, in each iteration of the game. When both operator $n$ and the WAP transmit in different power levels, they receive the following utilities,

$$W_{yx_n}^{(\mathrm{m})} = U_n^{(\mathrm{m})} \left( p_{x_n}^{(\mathrm{m})} \middle| p_y^{(\mathrm{w})}, \mathbf{p}_{-n}^{(\mathrm{m})} \right), \tag{11}$$

$$W_{yx_n}^{(\mathrm{w})} = U^{(\mathrm{w})} \left( p_y^{(\mathrm{w})} \middle| p_{x_n}^{(\mathrm{m})}, \mathbf{p}_{-n}^{(\mathrm{m})} \right), \tag{12}$$

$\forall y, x_n \in \{1, 2\}$. For a better understanding, we illustrate the utilities in Fig. 2, which is basically a $2 \times 2$ static game. According to the property of the utility functions, when operator $n$ increases its power level while the WAP keeps its transmit power unchanged, the utility function of operator $n$ increases and the utility function of the

**Fig. 2** Game analysis
between the operator $n$ and
the WAP in one iteration

| OPERATOR $n$ / WAP | $p_1^{(m)}$ | $p_2^{(m)}$ |
|---|---|---|
| $p_1^{(w)}$ | $W_{11}^{(w)}$ , $W_{11}^{(m)}$ | $W_{12}^{(w)}$ , $W_{12}^{(m)}$ |
| $p_2^{(w)}$ | $W_{21}^{(w)}$ , $W_{21}^{(m)}$ | $W_{22}^{(w)}$ , $W_{22}^{(m)}$ |

WAP decreases, and vice versa. Thus, we have

$$\begin{cases} W_{y2}^{(m)} > W_{y1}^{(m)}, & \forall y \in \{1,2\}; \\ W_{y2}^{(w)} < W_{y1}^{(w)}, & \forall y \in \{1,2\}; \\ W_{2x_n}^{(w)} > W_{1x_n}^{(w)}, & \forall x_n \in \{1,2\}; \\ W_{2x_n}^{(m)} < W_{1x_n}^{(m)}, & \forall x_n \in \{1,2\}. \end{cases} \tag{13}$$

Based on (13) above, $p^{(w)} = p_2^{(w)}$ and $p_n^{(m)} = p_2^{(m)}$ is the Nash equilibrium of the game. If $W_{22}^{(w)} + W_{22}^{(m)} > W_{11}^{(w)} + W_{11}^{(m)}, p^{(w)} = p_2^{(w)}$ and $p_n^{(m)} = p_2^{(m)}$ also achieve the Pareto optimality, which constitute the optimal strategies for both operator $n$ and the WAP.

However, if $W_{22}^{(w)} + W_{22}^{(m)} < W_{11}^{(w)} + W_{11}^{(m)}$, the game becomes a prisoner's dilemma where the social optimal point is not the Nash equilibrium solution. In order to achieve high and stable social welfare while guaranteeing the performance of the WU, we suppose the game is played in an iterated way. Thus, zero-determinant strategy can be applied by operator $n$ to unilaterally set a ratio relationship between the operator $n$ and the WAP, no matter what the strategy of the WAP is [22, 23]. Thus, when the WAP maximizes its individual utility, the social welfare can be optimized.

In the iterated game, as we do not consider the strategies of other operators, the strategy of the WAP can be defined as

$$q_{yx_ny'} = \sum_{\{x_{n'} | n' \in \mathcal{N} \setminus \{n\}\}} a_{yx_1x_2\ldots x_Ny_w}. \tag{14}$$

Thus, the transition matrix of the iterated process can be given as

$$\mathbf{H} = \begin{bmatrix} q_{111}z_{111} & q_{111}z_{112} & q_{112}z_{111} & q_{112}z_{112} \\ q_{121}z_{121} & q_{121}z_{122} & q_{122}z_{121} & q_{122}z_{122} \\ q_{211}z_{211} & q_{211}z_{212} & q_{212}z_{211} & q_{212}z_{212} \\ q_{221}z_{221} & q_{221}z_{222} & q_{222}z_{221} & q_{222}z_{222} \end{bmatrix}, \tag{15}$$

where $q_{yx_n1} + q_{yx_n2} = 1$ and $z_{yx_n1} + z_{yx_n2} = 1, \forall y, x_n \in \{1,2\}$.

In each iteration of the game, we assume the probability that the WAP transmits in power level $y$ while the operator $n$ transmits in power level $x_n$ is $d_{yx_n}$. Thus,

$$d_{yx_n} = v_y^{(w)} v_{x_n}^{(m)}, \tag{16}$$

$\forall y, x_n \in \{1, 2\}$. Denote $\mathbf{d} = [d_{11}, d_{12}, d_{21}, d_{22}]^\top$, we model the iterated process as a Markov chain. If

$$\mathbf{d}^\top \mathbf{H} = \mathbf{d}^\top, \tag{17}$$

can be established, the process achieves a stationary result. Define $\mathbf{H}' = \mathbf{H} - \mathbf{I}$, where $\mathbf{I}$ is the unit diagonal matrix. We then have

$$\mathbf{d}^\top \mathbf{H}' = \mathbf{0}. \tag{18}$$

Moreover, according to Cramer's rule, $\mathrm{adj}(\mathbf{H}')\mathbf{H}' = \det(\mathbf{H}')$, where $\mathrm{adj}(\mathbf{H}')$ is the adjugate matrix of $\mathbf{H}'$. Following the properties of the matrix determinant, we derive $\det(\mathbf{H}') = 0$. Thus,

$$\mathrm{adj}(\mathbf{H}')\mathbf{H}' = 0. \tag{19}$$

Based on (18) and (19), we deduce that each column of the $\mathrm{adj}(\mathbf{H}')$ is proportional to $\mathbf{d}^\top$. Accordingly, the dot product of $\mathbf{d}$ with any vector $\mathbf{f} = [f_1, f_2, f_3, f_4]^\top$ can be expressed as

$$\mathbf{d}^\top \cdot \mathbf{f} = \tag{20}$$

$$\det \begin{pmatrix} -1 + q_{111}z_{111} & -1 + q_{111} & -1 + z_{111} & f_1 \\ q_{121}z_{121} & -1 + q_{121} & z_{121} & f_2 \\ q_{211}z_{211} & q_{211} & -1 + z_{211} & f_3 \\ q_{221}z_{221} & q_{221} & z_{221} & f_4 \end{pmatrix},$$

where the second and third column of the determinant is only related to the strategies of the WAP and operator $n$, respectively. We set $\mathbf{z}_n = [-1 + z_{111}, z_{121}, -1 + z_{211}, z_{221}]^\top$ and $\mathbf{f} = \mathbf{W}^{(\mathrm{w})} - \beta \mathbf{W}_n^{(\mathrm{m})}$, where $\mathbf{W}^{(\mathrm{w})} = [W_{11}^{(\mathrm{w})}, W_{12}^{(\mathrm{w})}, W_{21}^{(\mathrm{w})}, W_{22}^{(\mathrm{w})}]$ and $\mathbf{W}_n^{(\mathrm{m})} = [W_{11}^{(\mathrm{m})}, W_{12}^{(\mathrm{m})}, W_{21}^{(\mathrm{m})}, W_{22}^{(\mathrm{m})}]$.

If

$$\mathbf{z} = \lambda \mathbf{f}, \tag{21}$$

we have

$$\mathbf{d}^\top \cdot \mathbf{f} = \mathbf{d}^\top \cdot \left( \mathbf{W}^{(\mathrm{w})} - \beta \mathbf{W}_n^{(\mathrm{m})} \right)$$
$$= F^{(\mathrm{w})} - \beta F_n^{(\mathrm{m})} = 0, \tag{22}$$

namely,

$$F_n^{(\mathrm{m})} = \frac{1}{\beta} F^{(\mathrm{w})}. \tag{23}$$

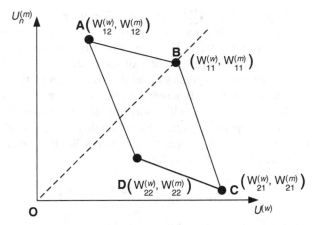

**Fig. 3** The utility of the operator $n$ vs the utility of the WAP when the operator $n$ adopts the zero-determinant strategy

where $F^{(w)}$ and $F_n^{(m)}$ are the expected utility of the WAP and operator $n$ in the $2 \times 2$ game, respectively.

Accordingly, the zero-determinant strategy for operator $n$ is calculated as,

$$\begin{cases} z_{111} = 1 + \lambda \left( W_{11}^{(w)} - \beta W_{11}^{(w)} \right), \\ z_{121} = \lambda \left( W_{12}^{(w)} - \beta W_{12}^{(w)} \right), \\ z_{211} = 1 + \lambda \left( W_{21}^{(w)} - \beta W_{21}^{(w)} \right), \\ z_{221} = \lambda \left( W_{22}^{(w)} - \beta W_{22}^{(w)} \right). \end{cases} \tag{24}$$

Moreover, as depicted in Fig. 3, the feasible region of the prisoner's dilemma is $ABCD$. The zero-determinant strategy of operator $n$ is characterized by a line starting at $O$ as shown in Fig. 3, i.e., as long as operator $n$ adopts the proposed zero-determinant strategy, no matter what the strategy of the WAP is, the final results of the game fall on one determined line [24]. In order to achieve the maximum utility for both operator $n$ and the WAP, operator $n$ should determine the line $OB$. Taking into account the constraint that $E_n^{(m)} \geq kE^{(w)}$, the value of $\beta$ satisfies

$$\frac{1}{\beta} = \max \left\{ k, \frac{W_{11}^{(m)}}{W_{11}^{(w)}} \right\}. \tag{25}$$

## 4.2 Game Analysis Among Operators

According to the analysis performed in the previous subsection, when the transmit powers of all other operators are fixed, the utility profiles of an operator $n \in \mathcal{N}$ as well as the WAP, namely, $\mathbf{W}_n^{(m)} = [W_{11}^{(m)}, W_{12}^{(m)}, W_{21}^{(m)}, W_{22}^{(m)}]$ and $\mathbf{W}^{(w)} =$

$[W_{11}^{(w)}, W_{12}^{(w)}, W_{21}^{(w)}, W_{22}^{(w)}]$, are fixed. Therefore, operator $n$ is able to configure the proposed zero-determinant strategy by setting a ratio between its own utility and the utility of the WAP. However, when the transmit powers from other operators are changed, the utility profiles of the operator $n$ and the WAP vary, and so does the game between operator $n$ and the WAP. That is, the behaviors of each operator will affect the utility functions of other operators. As each operator would like to increase its utility in a selfish way, we model the competitions among the operators as a non-cooperative game. The probability that each operator $n$ transmit in power level $x_n$, $\forall x_n \in \{1, 2\}$, $\forall n \in \mathcal{N}$, and the WAP transmits in power level $y$ can be expressed in the following form

$$\pi_{y x_1 \ldots x_N} = v_y^{(w)} \prod_{n \in \mathcal{N}} v_{x_n}^{(m), n}. \tag{26}$$

And inversely, it's straightforward to get

$$v_{x_n}^{(m), n} = \sum_{y, \{x_{n'} | n' \in \mathcal{N} \setminus \{n\}\}} \pi_{y x_1 \ldots x_N}, \tag{27}$$

$\forall x_n \in \{1, 2\}$, and

$$v_y^{(w)} = \sum_{\{x_n | n \in \mathcal{N}\}} \pi_{y x_1 \ldots x_N}, \tag{28}$$

$\forall y \in \{1, 2\}$.

Therefore, in the $2 \times 2$ game between each operator $n \in \mathcal{N}$ and the WAP, the probability of a situation that all other operators $n' \in \mathcal{N} \setminus \{n\}$ transmits in power level $x_{n'}$ is

$$\kappa_{x_1 \ldots x_{n-1} x_{n+1} \ldots x_N}^n = \sum_{y, x_n} \pi_{y x_1 \ldots x_N}, \tag{29}$$

$\forall x_n \in \{1, 2\}$.

In each situation, there is a corresponding utility profile for the $2 \times 2$ game between operator $n$ and the WAP. Following the game analysis in Sect. 4.1, we are able to obtain a stationary vector $\mathbf{d}(x_1, \ldots, x_{n-1}, x_{n+1}, \ldots, x_N)$ for each situation. Accordingly, we attain

$$\sum_{\{x_{n'} | n' \in \mathcal{N} \setminus \{n\}\}} \mathbf{T}_{x_1 \ldots x_{n-1} x_{n+1} \ldots x_N} = \boldsymbol{\Psi}_n, \quad \forall n \in \mathcal{N}, \tag{30}$$

where

$$\mathbf{T}_{x_1 \ldots x_{n-1} x_{n+1} \ldots x_N} =$$
$$= \kappa_{x_1 \ldots x_{n-1} x_{n+1} \ldots x_N}^n \mathbf{d}(x_1, \ldots, x_{n-1}, x_{n+1}, \ldots, x_N), \tag{31}$$

and

$$\boldsymbol{\Psi}_n = \left[ v_1^{(w)} v_1^{(m),n}, v_1^{(w)} v_2^{(m),n}, v_2^{(w)} v_1^{(m),n}, v_2^{(w)} v_2^{(m),n} \right]. \tag{32}$$

Moreover, based on the above definitions, we have

$$\begin{cases} \sum_{x_n=1}^2 v_{x_n}^{(m),n} = 1, \forall n \in \mathcal{N}; \\ \sum_{y=1}^2 v_y^{(w)} = 1. \end{cases} \tag{33}$$

Accordingly, when all the values of $\pi_{yx_1\ldots x_N}$ satisfy (30) and (33), all operators achieve a Nash equilibrium, where each operator cannot change its strategy unilaterally for a higher utility. Based on the value of $\pi_{yx_1\ldots x_N}$, each operator $n \in \mathcal{N}$ plays an $2 \times 2$ game with the WAP. The expected utility profile for operator $n$ is

$$\boldsymbol{\Omega}_n^{(m)} = \sum_{\{x_{n'}|n'\in\mathcal{N}\setminus\{n\}\}} \kappa_{x_1\ldots x_{n-1}x_{n+1}\ldots x_N}^n$$

$$\mathbf{W}_n^{(m)}(x_1,\ldots,x_{n-1},x_{n+1},\ldots,x_N). \tag{34}$$

The expected utility profile for the WAP is

$$\boldsymbol{\Omega}^{(w)} = \sum_{\{x_{n'}|n'\in\mathcal{N}\setminus\{n\}\}} \kappa_{x_1\ldots x_{n-1}x_{n+1}\ldots x_N}^n$$

$$\mathbf{W}^{(w)}(x_1,\ldots,x_{n-1},x_{n+1},\ldots,x_N). \tag{35}$$

Finally, the optimal zero-determinant power control strategy for operator $n$ is obtained as

$$\bar{\mathbf{z}}_n = \mathbf{z}_n \sum_{\{x_{n'}|n'\in\mathcal{N}\setminus\{n\}\}} \kappa_{x_1\ldots x_{n-1}x_{n+1}\ldots x_N}^n. \tag{36}$$

## 5  Simulation Results

In this section, we evaluate the performance of the operators and WAP with MATLAB. Without loss of generality, we assume that there are two operators trying to share the unlicensed spectrum with the WAP in a two-dimensional area. The operators are located at coordinates $(50, 0)$ and $(25, 43)$, and their scheduled MUs are located at coordinates $(90, 0)$ and $(-5, 43)$. The WAP is assumed to be located at the origin, and it serves a WU at coordinates $(0, 10)$. We assume two power levels for both the operators and the WAP, i.e., the power levels for both operators are, respectively, $\{600, 1200\}$ and $\{450, 900\}$. And the power levels for the WAP are chosen from $\{400, 800\}$. We set the price of unit transmit power for the WAP to be 0.001 and that for the WO to be 0.002. The power of the additive white noise is $\sigma = -105\,\mathrm{dBm}$.

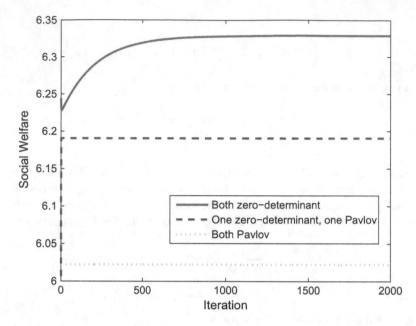

**Fig. 4** The social welfare vs. iteration when two WOs and the WAP share unlicensed spectrum at the same time

For better analysis, we compare our proposed zero-determinant strategy with the Pavlov strategy in the game. In the case of an operator choosing to implement the Pavlov strategy, if the received utility is higher than a predefined threshold, operator keeps the current transmit power level. If the received utility is smaller than the threshold, the operator switches to the other power level. Thus, the Pavlov strategy for an operator $n$ can be simply denoted by $\mathbf{z}_n = [1, 0, 0, 1]$, $\forall n \in \{1, 2\}$.

From the curves in Fig. 4, we discover that the social welfare of the game finally converges as the number of iterations increases. The converged value when both the operators adopt the proposed zero-determinant strategy is larger than the value when the first operator applies the proposed zero-determinant strategy and the second operator applies the Pavlov strategy. Moreover, the converged value when the first operator applies the proposed zero-determinant strategy and the second operator applies the Pavlov strategy is larger than the value when both the operators adopt the Pavlov strategy.

Furthermore, we evaluate the influence that the transmit power of the WAP can make to the system in Fig. 5. As the low power level of the WAP increases, we discover that the social welfare of the system gradually increases, but the increasing speed decreases. The reason behind this is that when the low power level of the WAP increases, the WU is able receive a higher data rate from the WAP. However, increasing the transmit power of the WAP also increases the interference to the operators coexisting in the unlicensed spectrum, which indicates the decrease in the increasing speed. We can also see from the plot that the social welfare when both

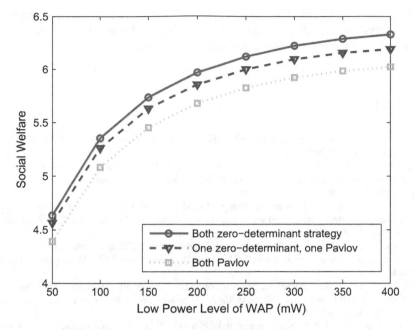

**Fig. 5** The social welfare vs. low power level of the WAP in the game

the operators adopt the proposed zero-determinant strategy is always larger than the social welfare when the first operator applies the proposed zero-determinant strategy and the second operator applies the Pavlov strategy. The social welfare when the first operator applies the proposed zero-determinant strategy and the second operator applies the Pavlov strategy is always larger than the social welfare when both the operators adopt the Pavlov strategy.

# 6  Summary

In this chapter, we formulate a layered power control game among all the operators and the WAP which jointly operate over a common unlicensed spectrum band. Each operator aims to maximize its own utility in a distributed manner with the protection of performance achieved by the WU in the Wi-Fi network. In the layered game, we first fix the transmit powers of all other operators and propose a zero-determinant strategy for the power control of each considered operator. The advantage of implementing the zero-determinant strategy is that operators can optimize the social welfare on their own, no matter what power control strategy is chosen by the WSP. To deal with the competition among the non-cooperative operators, we propose that each operator explores the predicted strategies from all other operators in all situations and hence determines its optimal zero-determinant strategy to reach the

Nash equilibrium results. The provided simulation results validate the correctness of the analysis in this chapter, and confirm that the high performance gain can be realized from the proposed zero-determinant strategies.

# References

1. "Cisco visual networking index: Global mobile data traffic forecast update, 2016–2021," San Jose, CA, USA, White Paper, Jan. 2017.
2. Q. Chen, G. Yu, H. Shan, A. Maaref, G. Y. Li and A. Huang, "Cellular meets WiFi: traffic offloading or resource sharing?" *Wireless Communications, IEEE Transactions,* vol. 15, no. 5, pp. 3354–3367, May 2016.
3. H. Zhang, X. Chu, W. Guo and S. Wang, "Coexistence of Wi-Fi and heterogeneous small cell networks sharing unlicensed spectrum," *Communications Magazine, IEEE,* vol. 53, no. 3, pp. 158–164, Mar. 2015.
4. J. Jeon, H. Niu, Q. C. Li, A. Papathanassiou, and G. Wu, "LTE in the unlicensed spectrum: evaluating coexistence mechanisms," *Globecom Workshops,* pp. 740–745, Austin, TX, Dec. 2014.
5. W. Xu, B. Li, Y. Xu, and J. Lin, "Lower-complexity power allocation for LTE-U systems: a successive cap-limited waterfilling method," *Vehicular Technology Conference, IEEE 81st,* Glasgow, UK, May 2015.
6. C. Cano, and D. J. Leith, "Coexistence of WiFi and LTE in unlicensed bands: A proportional fair allocation scheme," *Communication Workshop, IEEE International Conference,* pp. 2288–2293, London, UK, Jun. 2015.
7. H. Song, and X. Fang, "A spectrum etiquette protocol and interference coordination for LTE in unlicensed bands (LTE-U)," *Communication Workshop, IEEE International Conference,* pp. 2338–2343, London, UK, Jun. 2015.
8. N. Clemens, and C. Rose, "Intelligent power allocation strategies in an unlicensed spectrum," *New Frontiers in Dynamic Spectrum Access Networks, First IEEE International Symposium,* pp. 37–42, Nov. 2005.
9. F. S. Chaves, E. P. L. Almeida, R. D. Vieira, A. M. Cavalcante, F. M. Abinader, S. Choudhury, and K. Doppler, "LTE UL power control for the improvement of LTE/Wi-Fi coexistence," *Vehicular Technology Conference, IEEE 78th,* Las Vegas, NV, Sep. 2013.
10. C. Jiang, Y. Chen, K. J. R. Liu and Y. Ren, "Optimal pricing strategy for operators in cognitive femtocell networks," *Wireless Communications, IEEE Transactions on,* vol. 13, no. 9, pp. 5288–5301, Sep. 2014.
11. C. Jiang, Y. Chen, Y. Gao and K. J. R. Liu, "Joint spectrum sensing and access evolutionary game in cognitive radio networks," *Wireless Communications, IEEE Transactions on,* vol. 12, no. 5, pp. 2470–2483, May 2013.
12. C. Jiang, Y. Chen, Y. Yang, C. Wang and K. J. R. Liu, "Dynamic Chinese Restaurant game: theory and application to cognitive radio networks," *Wireless Communications, IEEE Transactions,* vol. 13, no. 4, pp. 1960–1973, Apr. 2014.
13. C. Jiang, Y. Chen, K. J. R. Liu and Y. Ren, "Renewal-theoretical dynamic spectrum access in cognitive radio network with unknown primary behavior," *Selected Areas in Communications, IEEE Journal,* vol. 31, no. 3, pp. 406–416, 2013.
14. X. Kang, R. Zhang and M. Motani, "Price-based resource allocation for spectrum-sharing femtocell networks: a stackelberg game approach," *Selected Areas in Communications, IEEE Journal,* vol. 30, no. 3, pp. 538–549, Apr. 2012.
15. S. Hajmohammad and H. Elbiaze, "Unlicensed spectrum splitting between femtocell and WiFi," in *Proc. IEEE ICC,* Budapest, Hungary, Jun. 2013.

16. S. Sagari, S. Baysting, D. Saha, I. Seskar, W. Trappe, and D. Raychaudhuri, "Coordinated dynamic spectrum management of LTE-U and Wi-Fi networks," in *Proc. IEEE DySPAN*, Stockholm, Sweden, Sep.–Oct. 2015.
17. F. Liu, E. Bala, E. Erkip, M. C. Beluri, and R. Yang, "Small-cell traffic balancing over licensed and unlicensed bands," *IEEE Trans. Veh. Technol.*, vol. 64, no. 12, pp. 5850–5865, Dec. 2015.
18. E. Jorswieck, L. Badia, T. Fahldieck, E. Karipidis, and J. Luo, "Spectrum sharing improves the network efficiency for cellular operators," *IEEE Commun. Mag.*, vol. 52, no. 3, pp. 129–136, Mar. 2014.
19. J. Panchal, R. Yates, and M. Buddhikot, "Mobile network resource sharing options: Performance comparisons," *IEEE Trans. Wireless Commun.*, vol. 12, no. 9, pp. 4470–4482, Sep. 2013.
20. "U-LTE: unlicensed spectrum utilization of LTE," *Huawei White Paper,* 2014.
21. Z. Han, D. Niyato, W. Saad, T. Basar, A. Hjorungnes, "Game Theory in Wireless and Communication Networks: Theory, Models and Applications," *Cambridge University Press,* 2011.
22. W. H. Press and F. J. Dyson, "Iterated prisoners' dilemma contains strategies that dominate any evolutionary opponent," *Proc. Natl Acad. Sci.*, vol. 109, no. 26, pp. 10409–10413, Apr. 2012.
23. H. Zhang, D. Niyato, L. Song, T. Jiang and Z. Han, "Zero-determinant strategy for resource sharing in wireless cooperations," *IEEE Trans. Wireless Commun.*, vol. 15, no. 3, pp. 2179–2192, Mar. 2016.
24. H. Zhang, D. Niyato, L. Song, T. Jiang and Z. Han, "Equilibrium analysis for zero-determinant strategy in resource management of wireless network," in *Proc. IEEE WCNC*, New Orleans, LA, Mar. 2015.

# Spectrum Matching in Unlicensed Band with User Mobility

## 1 Introduction

### 1.1 Coexistence Issue in Unlicensed LTE

Recent studies have highlighted that LTE technology has significant performance gains over Wi-Fi when operating in the unlicensed band [1]. The main advantages for unlicensed LTE (U-LTE) over Wi-Fi on the unlicensed spectrum include better link performance, medium access control, mobility management, and excellent coverage. These benefits have made U-LTE a promising technology. Due to the low power and high frequency transmission regulations imposed by Federal Communications Commission (FCC) on the unlicensed spectrum, small cell (SC) deployment in heterogeneous network (HetNet) is an ideal implementation scenario for the U-LTE. It is shown in [2] that U-LTE has a great potential in the ultra dense cloud SC deployment, which combines advantages of the cloud radio access network and ultra dense SCs. However, U-LTE is still in its infancy, and thus calls for great effort and careful design before meeting the requirements and regulations of both licensed and unlicensed transmissions. More specifically, how can we guarantee a fair coexistence of the newly joined cellular users (CUs) and the existing unlicensed users (UUs) on the unlicensed band? For traditional Wi-Fi transmission, which is collision avoidance based, UUs may back off to the co-channel U-LTE users if the interference level is above the energy detection threshold (e.g., $-62\,$dBm over $20\,$MHz) [1]. Thus without proper coexistence mechanisms, U-LTE transmissions can cause considerable interference on Wi-Fi transmissions. On the other hand, the interference from the co-channel Wi-Fi users may also degrade the U-LTE devices' performance, leading to the failure of meeting the quality of service (QoS) requirements for cellular transmissions. In addition, with limited unlicensed bands, U-LTE users need to compete with each other. Thus, there may exist inter-operator interference. To summarize, such unplanned and unmanaged deployment can result

© The Author(s) 2018
H. Zhang et al., *Resource Allocation in Unlicensed Long Term Evolution HetNets*,
SpringerBriefs in Electrical and Computer Engineering,
https://doi.org/10.1007/978-3-319-68312-6_4

in excessive interference to both Wi-Fi users and U-LTE users. Therefore, it is critical to design a coexistence mechanism to avoid such co-channel interference and guarantee the harmonious coexistence of Wi-Fi and LTE systems [3].

A fair coexistence is always evaluated from both the U-LTE and Wi-Fi users' point of view, and thus the coexisting interference can be generally categorized into three types: (1) the interference that CUs bring to the existing UUs; (2) the interference that the existing UUs bring to CUs; (3) the interference between multiple CUs who are reusing the same unlicensed band. Therefore, to satisfy these coexisting constraints, certain transmission restrictions should be imposed on both LTE and Wi-Fi systems. Some methods have been proposed to deal with the coexistence issues, for example the Channel Selection mechanism, Carrier-Sensing Adaptive Transmission (CSAT) and Opportunistic SDL [4]. The Channel Selection method enables the SCs to choose the cleanest channel based on the Wi-Fi and LTE measurements. When no clean channel is available, the CSAT algorithm can be used to apply adaptive TDM transmission based on the long-term carrier sensing of co-channel Wi-Fi activities. The SDL method allows to turn off the carrier aggregation when the SC is lightly loaded to avoid interference to Wi-Fi and transmission overheads. It is pointed out that, for most Wi-Fi and U-LTE SC deployments, Channel Selection is usually sufficient to meet the coexistence requirements [4]. In the case that one unlicensed band is the best choice for more than one CU, instead of allocating all such CUs to this unlicensed band, some CUs can be allocated to their second-best or third-best choices for more efficient network utilization. Thus, it becomes a critical issue, from the U-LTE SCs' perspective, that how to most efficiently allocate the unlicensed bands to multiple CUs so that the unlicensed resources can achieve the highest utilization while both cellular and Wi-Fi users' performances can meet their requirements/regulations.

## 1.2  Matching Theory for U-LTE

To find a proper solution for this unlicensed resource allocation problem between the CUs and coexisting UUs, we start by studying the features of the resource allocation problem and some existing solution methods. The future 5G mobile networks are expected to be characterized with features such as higher data rates, reduced end-to-end latency, better network coverage and so on. The heterogeneous characteristics exhibited by mobile users and the network density are the two major challenges that face the 5G design. Current architectures for mobile and cellular networks are highly centralized. The advantage of the centralized approach resides in its optimality, however with the gigantic information to be collected by the centralized agent (e.g., eNBs) and the extremely high computation complexity, the resulting service latency to the end users can be unsatisfying. In addition, considering the highly dynamic network environment, including the network topology change and channel condition varying, distributive network resource management is considered as a more efficient approach. More specifically, in the U-LTE context, with eNBs in control of the resource allocation, we can formulate the unlicensed resource allocation as a centralized optimization with interference constraints. As discussed previously, the network density, the user heterogeneity, the require for global information, as well as the mobility management, may result in high computation overhead and complexity, which make the centralized approach less efficient. As a popular mathematical tool, game theory is often used as an alternative approach to solve these problems in a distributed manner. We can model the resource allocation problem as the interactions between players under certain rules. However, game theory also has its limitations, for example that each player requires the knowledge of other players' actions in many cases, which restricts its distributive implementation. In addition, specific utility functions are always required for players, which is hard to realize in some practical applications.

Matching game can overcome some limitations of game theory and centralized optimization. It can model the competition and negotiation between the distinct user sets of LTE and Wi-Fi, and solve the problem in a semi-distributive way. We claim it as semi-distributive w.r.t. the fact that many operations in the matching algorithms are implemented distributively, including the information collection, preference list set up, local reject/accept decision making and so on, while certain operations may require global information from a centralized agent, such as the detection of a blocking pair. Different from the static resource allocation that has been studied [5, 6], which is a one-time allocation, the dynamic case is not a simple repeating of the static allocation over time. In this work, we propose a matching-based framework to tackle the dynamic U-LTE resource allocation problem, which explores the relations between the resource allocations of adjacent times. The major contributions are summarized as follows.

- We have summarized the coexistence issues of U-LTE into three categories. To solve such issues we have modeled the interactions between CUs and UUs as an interactive matching game: the stable marriage (SM) problem. The coexistence constraints are well interpreted through the set up of CUs' and UUs' preference lists.
- We have introduced two semi-distributed solutions: the Gale-Shapley (GS) algorithm and Random Path to Stability (RPTS) algorithm to tackle the resource allocations in U-LTE dynamically. Both mechanisms ensure network stability, while achieving relatively low computation complexity compared with the centralized optimization. Specifically, the proposed RPTS algorithm, which makes use of the relations between two time-adjacent matchings, further reduces complexity compared with GS, and is more suitable for dynamic networks.
- The external effect that occurs in many wireless resource allocation problems, which refers to instability caused by the inter-dependence of the matching players' preference lists, is addressed by the proposed Inter-Channel Cooperation (ICC) mechanism. The ICC procedure not only re-stabilize the system but also further improves network throughput.
- We evaluate the adaptability and robustness of the GS+ICC and RPTS+ICC mechanisms under two user mobility models: the Random Waypoint model, and the HotSpot model. The computation complexity and system optimality analysis are performed theoretically and also validated through simulations.

The rest of the chapter is organized as follows. The system model of the dynamic resource allocation in U-LTE is provided in Sect. 2. Then, the problem formulation and centralized solution are presented in Sect. 3. Due to the NP-hardness of the centralized solution, the semi-distributive matching approaches are introduced in Sect. 4. Two matching mechanisms are implemented in the time-independent way and the time-dependent way, respectively. Both theoretical and numerical analysis are provided in Sect. 5 to evaluate the proposed mechanisms. Finally, conclusion remarks are drawn in Sect. 6.

## 2   System Model

We consider a single carrier cellular network, where as illustrated in Fig. 1, there are $N$ CUs $\mathcal{CU} = \{cu_1, \ldots, cu_i, \ldots, cu_N\}$ subscribed to one cellular network operator (CNO). Each CU is served by its local eNB $\mathcal{BS} = \{bs_1, \ldots, bs_b, \ldots, bs_{B_1}\}$ with the allocated licensed spectrum. $B_1$ is the number of total eNBs. Due to the time varying traffic flow, some transmission requests can not be satisfied by the currently allocated licensed subband. We assume a set of such CUs travel around in the network with certain mobility patterns. Wherever CUs are located, they search for nearby UUs, and seek to share their unlicensed spectrum using the CA technique for supplemental downlink (SDL) transmission. We denote the set of

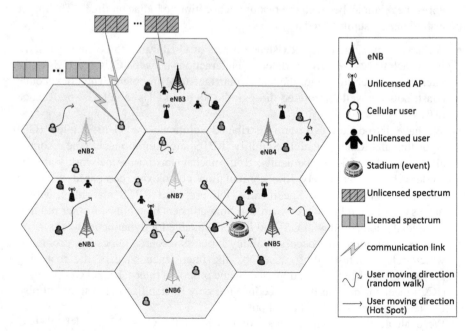

**Fig. 1**  System model

UUs as $\mathcal{UU} = \{uu_1, \ldots, uu_j, \ldots, uu_M\}$, and each UU is allocated with a specific unlicensed subband denoted as $\mathcal{F} = \{f_1, \ldots, f_j, \ldots, f_K\}$ for transmission. Typically, each unlicensed band is shared by multiple UUs based on the CSMA/CA regulation. All the pathless gains are independent of the unlicensed subbands, and fast fading is not considered in this work. All the unlicensed subbands use the same carrier frequency. To simplify the representation, we assume that $uu_j, uu_j \in \mathcal{UU}$ is assigned with the unlicensed band $f_k, f_k \in \mathcal{F}$. Each UU is served by its local Wi-Fi AP, denoted as $\mathcal{AP} = \{ap_1, \ldots, ap_j, \ldots, ap_{B_2}\}$, for transmitting/receiving data, where $B_2$ is the number of Wi-Fi Access Points (WAPs).

The pre-assigned licensed bands of CUs will be the primary carrier and will be aggregated with the shared unlicensed bands to enhance transmission. To access a clean unlicensed channel, CUs need to have the channel sensing phase before joining any unlicensed channel, and this channel sensing shall be repeated each time CUs joins a new unlicensed channel. During the channel sensing, CUs can detect the transmission energy on the target unlicensed channel and decide if this channel is clean or not by comparing with a threshold. The CUs then communicate with its local eNBs, who assist the CUs in accessing the unlicensed bands, through control signal exchanges using the pre-assigned licensed bands. On the other hand, to model the interference incurred at UUs from the sharing CUs, the locations of UUs and the Wi-Fi medium utilization (MU) estimation should be performed. The Wi-Fi MU monitoring is done by the WAPs through network listening, where all the U-LTE CUs are required to turn off the unlicensed spectrum sharing in this period. The Wi-Fi network listening decodes the preamble of any WiFi packet detected during this time and records its corresponding received signal strength indicator (RSSI), duration in µs (or NAV), modulation, coding scheme and source/destination address [7]. With the above estimated information of the unlicensed bands and the existing UUs, the WAPs will share with the U-LTE eNBs so that this information can be further shared with the CUs to select the proper partner UUs. To the best of our knowledge, there's no existing standard specifying how many unlicensed bands that each CU should use for aggregation in U-LTE, besides SDL is only considered as an enhancement to LTE transmission without any certain improvement guaranteed. Thus, without loss of generality, we assume in this work, that each CU will be matched to at most one UU, i.e., one unlicensed band. On the other hand, each unlicensed band can accommodate multiple CUs, depending on the number of its existing UUs.

As discussed in Sect. 1, the coexistence issues are categorized as follows: (1) the interference that CUs bring to the existing UUs; (2) the interference that the existing UUs bring to CUs; (3) The interference between multiple CUs who are reusing the same unlicensed band. We elaborate them one by one into the following constraints:

• It is well known that in Wi-Fi transmission, the UUs adopt the CSMA/CA mechanism for coexistence, which is different from the way that LTE system operates, who directly uses the spectrum without sensing. Thus, it is required that CUs should keep their interference incurred at the UUs to be sufficiently small, such that the channel is treated as "idle" by UUs. To achieve this requirement, we set the threshold of the any CU's interference as the energy level of UU's noise, denoted as $\sigma_{noise}$.

- On the other hand, not all unlicensed bands are clean enough for CUs to use. The existing UUs on some channels cause high interference that greatly reduces the transmission quality rather than enhancing the transmission. Thus, by restricting the signal to interference plus noise ratio (SINR) for $cu_i$ to be higher than the minimum requirement $\Gamma_i^{min}$ when choosing sharing UUs, we can guarantee CUs' QoS requirements.

- The inter-CU interference can be avoided by the management of eNBs. We assume the eNBs adopt TDMA for CUs who are sharing the same unlicensed bands, and each sharing CU is allocated an equal share of time. As more CUs are assigned to the same unlicensed band, each CU gets a smaller share of the resource. Thus, it might happen that, after assigned to some unlicensed channel, some CU may prefer to switch to another channel which has less CUs assigned. To avoid such situation, we design the ICC strategy to avoid the system-wide massive switching. The detailed mechanism will be discussed in Sect. 4.2.2.

## 3 Problem Formulation

There are majorly two factors that may cause network dynamics, one is the user mobility, and the other the channel fading. To model the network dynamics, which include the change of propagation gain, interference, and so on, we divide the simulation period $[0, T]$ into identical time slots $\Delta T$. The slot duration $\Delta T$ can be set according to specific applications. To precisely model the dynamic network due to user mobility, we can set $\Delta T$ to be sufficiently small that during each time slot $(t, t + 1), \forall t \in \{1, \ldots, t, \ldots, T\}$, the user distribution and channel conditions can be treated as static. In other words, we assume that the resource allocation only happens at the beginning of each time slot. Thus, the formulation of our dynamic resource allocation problem will be built based on each specific time slot $(t, t + 1)$.

In order to pursuit higher spectrum efficiency, we allow multiple CUs to share the same unlicensed channel as long as the incurred coexisting interference is acceptable for each co-channel CU and UU. Each CU is only allowed to be allocated to one unlicensed channel. In other words, it is a many-to-one matching between CUs and the unlicensed bands (i.e., UUs). To model the dynamic resource allocation problem between CUs and UUs, we adopt a binary matrix for each time slot, denoted as $\rho(t) = \{\rho_{i,j} | cu_i \in \mathcal{CU}, uu_j \in \mathcal{UU}\}$. $\rho_{i,j}(t)$ is a binary value equal to 1 or 0 indicating if $cu_i$ is or is not assigned with $uu_j$ (i.e., subband $f_j$) at time $t$. To dynamically maximize the social welfare, we endeavor to find the allocation matrix $\rho(t)$ sequentially at each time that can achieve the highest overall performance of CUs and UUs.

## 3.1   CUs' Performance

In this work, we assume that U-LTE for CUs' SDL transmission. Thus, $cu_i$ is the receiver and its local eNB $bs_b$ is the transmitter. The interference from the coexisting UU is also incurred on the receiver $cu_i$. Thus, The received SINR at $bs_b$ when sharing $f_j$ with $uu_j$ at time $t$, used to measure the performance $cu_i$, is represented as follows:

$$\Gamma_{i,j}(t) = \frac{\rho_{i,j}(t)P_{b,i}(t)g_{b,i}(t)}{\sigma_N^l + P_{j,i}(t)h_{j,i}(t)}, \tag{1}$$

where $P_{b,i}(t)$ and $g_{b,i}(t)$ are the transmission power and channel gain from $bs_b$ to $cu_i$ at time $t$, respectively. $P_{j,i}(t)$ and $h_{j,i}(t)$ represent the transmission power and channel gain from $uu_j$ to $cu_i$, respectively. $\sigma_N^l$ is the licensed channel noise.

## 3.2   UUs' Performance

On the other hand, UUs will be interfered by the spectrum sharing from CUs, although the interference is controlled to be small. In the case that $f_j$ is utilized by $cu_i$ for UL transmission, $uu_j$ is interfered by the transmitter $cu_i$'s power. Thus, the interference of $uu_j$ from $cu_i$ at time $t$ is denoted as follows:

$$\text{Intf}_{i,j}^{UL}(t) = P_{i,j}(t)h_{i,j}(t), \tag{2}$$

where $P_{i,j}(t)$ and $h_{i,j}(t)$ represent the transmission power and channel gain from $cu_i$ to $uu_j$, respectively.

While $f_j$ is utilized by $cu_i$ for DL transmission, $uu_j$ is interfered by the transmitter $bs_b$'s transmission power. Thus, the interference of $uu_j$ from $bs_b$ at time $t$ is denoted as follows:

$$\text{Intf}_{i,j}^{DL}(t) = P_{b,j}(t)h_{b,j}(t), \tag{3}$$

where $P_{i,j}(t)$ and $h_{i,j}(t)$ represent the transmission power and channel gain from $bs_b$ to $uu_j$, respectively.

Thus, $uu_j$'s received interference $\text{Intf}_{i,j}$ equals to $P_{i,j}(t)h_{i,j}(t)$ if $cu_i$ is a transmitter, and $\text{Intf}_{i,j} = P_{b,j}(t)h_{b,j}(t)$ if $cu_i$ is a receiver. We represent $uu_j$'s SINR at time $t$ when sharing $f_j$ with $cu_i$ as:

$$\Gamma_{j,i}^{UU}(t) = \frac{\mu_{i,j}(t)P_j(t)g_j(t)}{\sigma_N^u + \text{Intf}_{i,j}}, \tag{4}$$

where $P_j(t)$ and $g_j(t)$ is the transmission power and channel gain for $uu_j$, respectively. $\sigma_N^u$ is the unlicensed spectrum noise.

Now, we formulate the dynamic spectrum sharing problem in U-LTE as a sequence of static resource allocation problems for each time slot. With the objective of dynamically maximizing the system throughput, the problem formulation is shown as follows:

$$\max_{\rho_{i,j}(t)} \sum_i (\sum_j \frac{\rho_{i,j}(t)}{\sum_i \rho_{i,j}(t)} f_k \log(1 + \Gamma_{i,j}^{CU}(t)))$$

$$+ \sum_j (\sum_i \frac{1}{\sum_i \rho_{i,j}(t)} f_k \log(1 + \Gamma_{j,i}^{UU}(t))), \tag{5}$$

**s.t. :**

$$\Gamma_{i,j}^{CU}(t) \geq \Gamma_i^{min}, \forall cu_i \in \mathcal{CU}, \tag{6}$$

$$\text{Intf}_{i,j}(t) \leq \sigma_{\text{noise}}, \forall uu_j \in \mathcal{UU}, \tag{7}$$

$$\sum_j \rho_{i,j}(t) \leq 1, \forall cu_i \in \mathcal{CU}, \tag{8}$$

$$\sum_i \rho_{i,j}(t) \leq 1, \forall uu_j \in \mathcal{UU}, \tag{9}$$

Notice that for any $uu_j$, its associated unlicensed band is pre-assigned, and is denoted as $f_k, \forall f_k \in \mathcal{F}$. Equation (6) is the SINR requirement that any CU should satisfy if to reuse a certain unlicensed band. It limits the interference CU receives from the coexisting UUs on the unlicensed band. Equation (7) represents the maximum interference that any UU can allow resulting from the coexisting CUs on the unlicensed band to guarantee fair coexistence. Equations (8) and (9) are the capacity requirements for CUs and UUs. Each CU can be allocated to only one UU (i.e., one unlicensed band), and each UU is only allowed to matched to one CU.

The formulated problem becomes a sequential mix integer nonlinear programming (MINLP) problems, which are in general NP-hard to solve centrally [8]. In addition, to cope with network dynamics, distributive solutions usually act more quickly with lower computation complexities. Thus, we introduce the matching-based approach as the semi-distributive solution, which will be discussed in the following section.

## 4  Dynamic Matching Framework

Matching theory, as a mathematical framework attempting to describe the formation of mutually beneficial relations, has been successfully applied to many economic fields. Recently, it has emerged as a promising technique for future wireless resource allocation solutions, which overcomes some limitations of traditional game theory and centralized optimization [9]. The advantages of matching theory include suitable models for various communication issues, preference interpretations for system constraints and efficient algorithms for desired objectives. As a fundamental requirement for wireless systems, the concept of stability should be treated with great attention. Generally speaking, the stability notion in wireless resource allocation applications refers to the situation where no player pairs/groups (e.g., CU and UU pairs) have the incentive to violate the current assignment under the table for their own benefits. The instability caused by such deviations is undesirable in any communication systems. To give a general idea of how matching theory works, we take the classical matching model *stable marriage* (SM) [10] as an example. Assume there are a set of men and a set of women, each of which is called a matching agent. A *preference list* for each agent is an ordered list based on the preferences over the other set of agents who he/she finds acceptable. A matching consists of (man, woman) pairs. A basic requirement, the *stability* concept refers to the case that, in a matching there exists no (man, woman) pair, who both have the incentive to leave their current partners and form a new marriage with each other.

The formulated optimization problem in Sect. 3, looking from a matching point of view, can be modeled as a one-to-one matching game between the CUs and UUs, which resulting in a many-to-one matching between the CUs and unlicensed bands. Typically, the two-sided one-to-one matching problem has been well studied using the SM model we discussed previously. Different from the traditional SM model, the sequential optimization problems correspond to a dynamic many-to-one matching problem. Intuitively, we can tackle the sequential optimization problems by taking each individual time interval as a traditional SM game, and solving each of them independently over time. This idea will be elaborated in Sect. 4.2. However, in a dynamic network, both the network topology and channel conditions are not isolated in time, and thus there exists some relations between the resource allocations for adjacent times. Instead of solving the optimization problem independently, we may explore the relation between any two time-adjacent networks, and make use of it for the resource allocation. Under such belief, we propose another matching approach, called the random path to stability (RPTS) algorithm, to address the network dynamics. By taking advantage of the relations over time, we can lower the solution cost as compared to the repeated GS approach. The second approach will be discussed in more details in Sect. 4.3. A detailed implementation for both approaches is shown in Fig. 2.

## 4.1  Basics of the SM Game

The SM problem is a bipartite matching problem with two-sided preferences. We assume an instance $I$ of the SM problem, which involves a set of men $\mathcal{M} = \{m_1, \ldots, m_{n1}\}$ and a set of women $\mathcal{W} = \{w_1, \ldots, w_{n2}\}$. Each man ranks the women from the most favorite to the least favorite based on his preferences, such as personalities, interests, income and so on. Such ranking is called men's preference list. On the other hand, women do the same thing to men. Once the preference lists

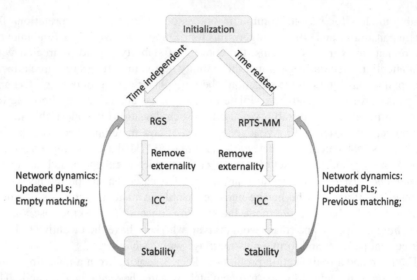

**Fig. 2** Matching implementations

are built, the players (men/women) take actions according to the lists. Each man or woman is allowed to be matched to at most one partner. The final result of this SM matching consists of man-woman pairs, while the objective of the matching diverges. The stability definition for the SM instance is provided in Definition 1.

**Definition 1** Let $I$ be an instance of SM, and $\mathcal{M}$ be a matching in $I$. A pair $(m_i, w_j)$ blocks $M$, or is a blocking pair of $M$, if the following conditions are satisfied relative to $M$:

(1) $m_i$ is unassigned or prefers $w_j$ to $\mathcal{M}(m_i)$;
(2) $w_j$ is unassigned or prefers $m_i$ to $\mathcal{M}(w_j)$.

$\mathcal{M}$ is said to be stable if it admits no blocking pair.
$\mathcal{M}(x)$ refers to the partner of $x$ in $\mathcal{M}$, and $x$ can be either a man or a woman.

Similar to the SM matching game, we assume CUs to be men and UUs to be women. Then as the pre-procedure of all matching algorithms, we first establish each player's preference list over the other set of players. With the channel sensing results from both CUs and WAPs, CUs and UUs can set up their preference lists. Pay attention that, UUs' preference lists set up are not actually performed by UUs, but by U-LTE eNBs and then update to all CUs. More specifically, combining the Wi-Fi MU information from WAPs and the CUs' channel sensing results, the U-LTE eNBs are able to generate the UUs' preference lists representing the interests of UUs. Thus, the interaction between the CUs and UUs are in fact interactions between CUs and the LTE-U eNBs. The preference of a CU $cu_i, cu_i \in \mathcal{CU}$ over its neighboring UUs $uu_j, uu_j \in \mathcal{UU}$ is based on $cu_i$'s achievable transmission rate when $uu_j$'s unlicensed spectrum $f_j$. Notice here, that each unlicensed band could be shared with multiple UUs as long as such UUs satisfy the unlicensed transmission regulation. Thus, each unlicensed band can also be shared within multiple CUs, which brings interference between coexisting CUs. However, before CUs join any unlicensed spectrum, they have no idea on the other coexisting CUs. Thus, the preference of $cu_i$ over $uu_j$ (on $f_j$) at time $t$ is simply assumed to be $cu_i$'s transmission rate when only itself is sharing $f_j$, and is represented as follows:

$$\mathcal{PL}_{i,j}^{CU}(t) = f_j \log(1 + \Gamma_{i,j}^{CU}(t)). \tag{10}$$

On the other hand, the preferences of $uu_j$ over $cu_i$ at time $t$ is based on $uu_j$'s achievable transmission rate when sharing spectrum with $cu_i$, which is shown as follows,

$$\mathcal{PL}_{j,i}^{UU}(t) = f_j \log(1 + \Gamma_{j,i}^{UU}(t)). \tag{11}$$

## 4.2  Time-Independent Implementation

### 4.2.1  The GS Algorithm

Generally speaking, a stable matching for an SM instance can be achieved by using the GS algorithm. A stable matching is always guaranteed by using the GS algorithm as stated in Theorem 1 [11].

**Theorem 1** *Given an instance of SM, the GS algorithm constructs in $\mathcal{O}(m)$ time, the unique man-optimal stable matching, where m is the number of acceptable man-woman pairs.*

The GS algorithm consists of sequential proposing and accepting/rejecting actions. Each iteration starts with the men proposing to the most favorite women (the first entity on the preference list) on their current preference lists. After proposing, the women being proposed to are removed from the men' preference lists. Then the women decide whether to accept or reject the proposals they've received so far based on their preference lists over the men. If the cumulative proposals exceed 1, each woman chooses to keep the man that she favors most, and rejects the rest. This proposing and accepting/rejecting iteration runs for as many rounds as needed until all men matched or all men preferences are empty, and its convergence is provided in [10]. The implementation details of the modified GS in U-LTE can be found in Algorithm 1.

### 4.2.2  Eliminating the External Effect

Notice here, for the conventional SM game, a stable matching is guaranteed using the GS algorithm. However, this conclusion is only correct under the canonical matching assumption, which implies that the preference of any player depends

---

**Algorithm 1** Man-oriented GS (GS) algorithm

---

**Input:**$\mathcal{CU}$, $\mathcal{UU}$, $\mathcal{PL}^{CU}(t)$, $\mathcal{PL}^{UU}(t)$, $q$
**Output:**Matching $\mathcal{M}(t)$

   Construct the set of unmatched $\mathcal{CU}_{un}$, set $\mathcal{CU}_{un} = \mathcal{CU}$;
   **while** $\mathcal{CU}_{un} \neq \emptyset$ and $\mathcal{PL}^{CU} \neq \emptyset$ **do**
      **CUs proposal to UUs;**
      **for all** $cu_i \in \mathcal{CU}_{un}$ **do**
         Propose to the first UU it in its preference list $uu_j$, and remove $uu_j$ from $\mathcal{PL}^{UU}$;
      **end for**
      **UUs make decisions;**
      **for all** $uu_j \in \mathcal{UU}$ **do**
         **if** $uu_j$ has received proposals no more than 1 **then**
            $uu_j$ keeps the proposal, and remove this CU from $\mathcal{CU}_{un}$;
         **else**
            $uu_j$ keeps the most preferred proposal, and rejects the rest;
            Remove this favorite CU from the $\mathcal{CU}_{un}$, and add the rejected CUs into the $\mathcal{CU}_{un}$;
         **end if**
      **end for**
   **end while**

---

solely on the local information about the other type of players. However, for the case, where players' preferences are affected by other players' choices/decision, the matching resulting from the traditional GS algorithm no longer guarantees stability. Any matching with the inter-dependence of players' preferences, is called matching with externality [12]. In fact, the external effect is commonly seen in the wireless resource allocation problems due to users' coexistence interference. Unfortunately, our proposed framework also exists externality, since CUs' performances are indeed affected by the other CUs' choices. For example, if too many CUs are matched to the same unlicensed band, then each of them will be assigned a smaller share (by TDMA) than they expect in the preference list, in which case some CU may have the incentive to change to some unlicensed band (i.e., a different UU) that is not assigned any CU or assigned with less CUs. In addition, each CU is only admitted by its matched UU, but are not necessarily acceptable to the coexisting UUs on the same unlicensed band, and vice verse for the UUs on the other admitted CUs. Such many-to-one relationship between CUs and unlicensed bands brings externality in the channel allocation, thus making the resulting matching no longer stable nor valid.

In order to eliminate such externality, we propose the Inter-Channel Cooperation (ICC) strategy to validate and re-stabilize the matching. As a first step, those invalid sharing, i.e., if a CU is not admitted by at least one of the UUs on the allocated unlicensed band, should be forbidden or removed. As we have discussed before, eNBs are representing the UUs/unlicensed bands to interact with CUs, thus after the matching using GS, eNBs can help find out those invalid CUs/UUs. Then, such invalid sharing are removed by eNBs informing both the related CU and UU, and also help update their preferences by removing invalid players from the lists. Such invalid sharing detection requires centralized information and operation, i.e., the assistance of eNBs. The next step, is to re-stabilize the matching. Pay attention that, since UUs are not really involved in the interaction, but represented by eNBs, thus, the whole matching is based on the interest of the CUs. As long as the unlicensed transmission regulation are meet, the allocation strategy should focus on how to further improve CUs' performances. Therefore, at this time point, the external effect can be evaluated from the CUs' perspective only. In other words, it becomes a one-sided "stability" problem. The new "stability", different from Definition 1, relies on the equilibrium among all CUs (i.e., there is no CU that has incentive to make any change). We call this one-sided "stability" as "Pareto Optimality" in matching theory [11]. The definition of Pareto optimal is given as follows.

**Definition 2** Pareto Optimal: A matching is said to be Pareto Optimal if there is no other matching in which some player (i.e., CU) is better off, whilst no player is worse off.

Accordingly, we provide the new definition of BP for the one-sided matching problems in Definition 3.

**Definition 3** A BP in the one-sided matching: A CU pair $(cu_i, cu_j)$ is defined as a BP, if both $cu_i$ and $cu_j$ are better off after exchanging their partners.

The basic idea of ICC is described as follows: firstly remove all invalid (CU, UU) pairs. The removed CUs will remain unmatched during the rest of the ICC algorithm. This is because ICC is designed based on the Pareto optimality, which is the one-sided stability. If any (CU, CU) pair would exchange partners, both of the CUs must agree with the exchange (i.e., benefit from the exchange). Now that the invalid (CU, UU) pairs have been removed, meaning these CUs currently have no UU partners (i.c., unlicensed resource), then it is reasonable that no other CU is willing to exchange partner (i.e., unlicensed resource) with such CUs. The second step is to search all "unstable" CU-CU pairs (who have the exchange incentive) regarding the current matching; secondly, check whether the exchange between such a pair is allowed (beneficial to related CUs); thirdly find the allowed pair, which provides the greatest throughput improvement, switch their partners, and update the current matching; then keep searching "unstable" CU-CU pairs until the trade-in-free environment is reached. The detailed ICC algorithm is illustrated in Algorithm 2.

In Algorithm 2, we transform the current matching $\mathcal{M}$ (i.e., $\mathcal{M}(t)$ generated by GS) into $\mathcal{M}'$. We define $\mathcal{M}(cu_{i1}) = uu_{j1}$, $\mathcal{M}(cu_{i2}) = vu_{j2}$. The utility of $cu_i$ is represented as $U(cu_i) = f_j \log(1 + \Gamma_{i,j}^{CU})$, and $\Delta U(cu_i) = U(cu_i)' - U(cu_i)$, where $U(cu_i)'$ is the utility after exchanging partner with another CU. The optimal BP is defined in (12).

$$(cu_{i1}^*, cu_{i2}^*) = \underset{(cu_{i1}, cu_{i2})}{\text{argmax}} \sum_{cu_{i1} \in \mathcal{M}_t(uu_{j1})} \Delta U(cu_{i1}) + \sum_{cu_{i2} \in \mathcal{M}_t(uu_{j2})} \Delta U(cu_{i2}), \quad (12)$$

where the CU pair $(cu_{i1}, cu_{i2})$ is allowed to exchange partners. The convergence of ICC is guaranteed by the irreversibility of each switch. The dynamic stability, under the time-related implementation, is reached by adopting the GS+ICC algorithm iteratively in each time slot.

---

**Algorithm 2** Inter-Channel Cooperation (ICC) strategy

---

**Input:** Existing matching $\mathcal{M}$, updated preference lists $\mathcal{PL}^{CU}(t)$ w.r.t. $\mathcal{M}$;
**Output:** Stable matching $\mathcal{M}'$.

 1: $\mathcal{M}' = \mathcal{M}$;
 2: Remove all invalid (CU, UU) pairs;
 3: **while** $\mathcal{M}'$ is not Pareto optimal **do**
 4:     Search the set of "unstable" CU-CU pairs $\mathcal{BP}(t)$ based on $\mathcal{PL}^{CU}(t)$;
 5:     **for all** $(cu_{i1}, cu_{i2}) \in \mathcal{BP}(t)$ **do**
 6:         **if** $\exists cu \in \mathcal{M}'(uu_{j1}^{k1}) \cup \mathcal{M}'(uu_{j2}^{k2})$, $\Delta U(cu) < 0$ **then**
 7:             $(cu_{i1}, cu_{i2})$ are not allowed to exchange partners;
 8:         **else**
 9:             $(cu_{i1}, cu_{i2})$ are allowed to exchange partners;
10:         **end if**
11:     **end for**
12:     Find the optimal BP $(cu_{i1}^*, cu_{i2}^*)$;
13:     $cu_{i1}^*$ and $cu_{i2}^*$ switch partners;
14:     $\mathcal{M}' \leftarrow \mathcal{M}'/\{(cu_{i1}^*, \mathcal{M}'(cu_{i1}^*)), (cu_{i2}^*, \mathcal{M}'(cu_{i2}^*))\}$;
15:     $\mathcal{M}' \leftarrow \mathcal{M}' \cup \{(cu_{i1}^*, \mathcal{M}'(cu_{i2}^*)), (cu_{i2}^*, \mathcal{M}'(cu_{i1}^*))\}$;
16:     Update $\mathcal{PL}^{CU}(t)$ based on $\mathcal{M}'$;
17: **end while**

---

## 4.3   Time-Dependent Implementation

Although we can use GS+ICC repeatedly in each time slot to find stable solutions, it is not computationally efficient to do so. Let's consider the case that for two adjacent time slots, the network condition varies very slightly. In other words, only a small number of users' preferences are changed. Under such small network variation, the stable matching also only varies a little bit regarding a small number of players. Thus, instead of redoing the whole matching, we can utilize the relations between the matching of the current time slot and that of the previous slot to transform the previously unstable matching into stable again. There, in this section, we propose an adaptive matching approach: the random path to stability (RPTS), also called the Roth Vanda-Vate (RVV) Algorithm [13]. The basic idea of RPTS mechanism is to use divorce and remarriage operations to transform an existing matching into stable again. Based on the previous matching $\mathcal{M}(t-1)$ at time $t-1$ and the updated preference lists $\mathcal{PL}^{CU}(t)$, $\mathcal{PL}^{UU}(t)$ at time $t$, RPTS algorithm provides a stable matching $\mathcal{M}(t)$ at time $t$.

For a SM instance $I$, consisting of the men set $\mathcal{CU}$ and women set $\mathcal{UU}$. As shown in Algorithm 3, the RPTS algorithm starts from an initial matching $M_0$, which is the previous matching $\mathcal{M}(t-1)$ of time $t-1$,[1] and finally terminates with a stable matching $\mathcal{M}(t)$ at time $t$. Each loop of RPTS comes with a matching $\mathcal{M}_i$, and finally terminates with a stable matching. A set $A$ is utilized during the loop iteration of RPTS, which is initially empty. $\mathcal{M}_i|_A$ denotes $\mathcal{M}_i \cap (A \times A)$, and $I|_A$ denotes the

---

[1]We assume the initial matching $M_0$ to be empty.

---

**Algorithm 3** Random path to stability (RPTS) algorithm

---

**Input:** Stable matching $\mathcal{M}(t-1)$ in the previous time $t-1$
**Output:** Stable matching $\mathcal{M}(t)$ at time $t$
1: **Initialization:**
2: $\mathcal{M}_i = \mathcal{M}(t-1), A = \emptyset$;
3: **while** $\mathcal{M}(t)$ is not stable in $\mathcal{I}$ **do**
4:     **if** There exists $(a_i, b_j) \in bp(I, \mathcal{M}_i)$ such that $a_i \notin A$, and $b_j \in A$ **then**
5:         *add* $a_i$;
6:     **else**
7:         choose $(m_i, w_j) \in bp(I, \mathcal{M}_i)$;
8:         *satisfy* $(m_i, w_j)$;
9:     **end if**
10: **end while**
11: $\mathcal{M}(t) = \mathcal{M}_i$

---

**Algorithm 4** *add* procedure for RPTS algorithm

---

**Input:** $a_i, \mathcal{M}_i$
**Output:** $A, \mathcal{M}_i$
1: **if** $a_i$ is assigned in $\mathcal{M}_i$ **then**
2:     $\mathcal{M}_i = \mathcal{M}_i / \{(a_i, \mathcal{M}_i(a_i))\}$;
3: **end if**
4: $A = A \cup \{a_i\}$;
5: **while** $a_i$ is blocking agent in $(I|_A, \mathcal{M}_i|_A)$ **do**
6:     $a_i$ is the proposer;
7:     $(a_i, b_i) \doteq bestbp(I|_A, \mathcal{M}_i|_A, a_i)$;
8:     $a_z \doteq a_i$;
9:     **if** $b_i$ is assigned in $\mathcal{M}_i$ **then**
10:         $\mathcal{M}_i = \mathcal{M}_i / \{(\mathcal{M}_i(b_i), b_i)\}$;
11:         $a_i = \mathcal{M}_i(b_i)$;
12:     **end if**
13:     $\mathcal{M}_i = \mathcal{M}_i \cup \{(a_z, b_i)\}$;
14: **end while**

---

sub-instance of $I$ obtained by deleting every member of $(\mathcal{CU} \cup \mathcal{UU})/A$, including the preference lists. The loop in RPTS iterates as long as $\mathcal{M}_i$ is not stable in $I$. During each iteration, if there's a blocking pair $(a_i, b_j)$ in such that $a_i \notin A$ and $b_j \in A$, procedure *add* is called with parameter $a_i$. Otherwise, the *satisfy* procedure is called with parameters $a_i$ and $b_j$ ($a_i \notin A, b_j \notin A$). Notice here, $a_i$ can be either a man or a woman, and similarly for $b_j$. The two procedures *add* and *satisfy* are maintained to ensure: (1) no member of $A$ is assigned in $\mathcal{M}_i$ to a member outside of $A$; (2) $\mathcal{M}_i|_A$ is stable in $I|_A$.

In the *add* procedure, $a_i$ is either a man or a woman, which doesn't belong to $A$. Our task is to ensure that upon the arrival of $u_i$, the matching can be restabilize so that $\mathcal{M}_i|_A$ is again stable in $I|_A$. We start by divorcing the pair $(a_i, \mathcal{M}_i(a_i))$ if $a_i$ is assigned in $\mathcal{M}_i$, and add $a_i$ to the set $A$. If $a_i$, as the current proposer, is a blocking agent (i.e., involved in a blocking pair) in $(I|_A, \mathcal{M}_i|_A)$, we search the best blocking pair $(a_i, b_i)$ in $(I|_A, \mathcal{M}_i|_A)$ w.r.t. $a_i$'s preference list. This $b_i$ must belong to $A$, and

---

**Algorithm 5** *satisfy* procedure for RPTS algorithm

---

**Input:** $(m_i, w_j), \mathcal{M}_i$
**Output:** $A, \mathcal{M}_i$
  1: $A = A \cup \{(m_i, w_j)\}$;
  2: **if** $m_i$ is assigned in $\mathcal{M}_i$ **then**
  3:      $\mathcal{M}_i = \mathcal{M}_i/\{(m_i, \mathcal{M}_i(m_i))\}$;
  4: **end if**
  5: **if** $w_j$ is assigned in $\mathcal{M}_i$ **then**
  6:      $\mathcal{M}_i = \mathcal{M}_i/\{(\mathcal{M}_i(w_j), w_j)\}$;
  7: **end if**
  8: $\mathcal{M}_i = \mathcal{M}_i \cup \{(m_i, w_j)\}$;

---

will be divorced from $\mathcal{M}_i(b_i)$ if it's assigned in $\mathcal{M}_i$. Then this $\mathcal{M}_i(b_i)$ becomes the next proposer, and we can marry $(a_i, b_i)$ in $\mathcal{M}_i$. The while loop continues as long as the current proposer is a blocking agent in $(I|_A, \mathcal{M}_i|_A)$.

In the *satisfy* procedure, $a_i \notin A$ and $b_j \in A$, and we assume $a_i, b_j$ to be $m_i, w_j$. Our task is to satisfy both $w_i$ and $w_j$. We start by adding $w_i$ and $w_j$ to $A$. If $m_i/w_j$ is assigned in $\mathcal{M}_i$, we divorce it from its partner $\mathcal{M}_i(m_i)/\mathcal{M}_i(w_j)$. Their partners (if any) will remain unassigned. Then we add this blocking pair $(m_i, w_j)$ to $\mathcal{M}_i$.

The dynamic stability, under the time-dependent implementation, is reached by adopting the RPTS+ICC algorithm iteratively. Regarding the convergence of RPTS mechanism in the SM model, a conclusion is stated in Theorem 2 [13], and the proof is provided as follows.

**Theorem 2** *Let $\mathcal{M}_0$ be an arbitrary matching for a SM instance I with N men and M women. Then there exists a finite sequence of matchings $\mathcal{M}_0, \ldots, \mathcal{M}_s$, where $\mathcal{M}_i$ is stable, and for each $1 \le i \le s$, $\mathcal{M}_i$ is obtained from $\mathcal{M}_{i-1}$ by satisfying a blocking pair of $\mathcal{M}_{i-1}$. Moreover, $\mathcal{M}_s$ can be obtained in $\mathcal{O}((N + M)m)$ overall time, where m is the number of acceptable man-woman pairs in I.*

*Proof* During each iteration of RPTS, $A$ increases in size by either one (*add* procedure) or two elements (*satisfy* procedure). At the end of each such iteration, we have $\mathcal{M}_i|_A$ is stable in $I|_A$. Hence we are bound to ultimately reach the outcome that $\mathcal{M}_s$ is stable in $I$ (when $A$ increase to the size of $(N + M)$, in which case RPTS terminates.

The complexity of RPTS is obtained by observing that $A$ increases in size by a minimum number of one element at each loop iteration of RPTS. Since $A \le (N + M)$, it follows that the same upper bound applies to the number of execution of RPTS. Each proposal-rejection sequence during an execution of *add* involves at most $m$ pair of agents. Thus, each iteration of *add* runs in $\mathcal{O}(m)$ time. While each call of the *satisfy* procedure takes $\mathcal{O}(1)$ time (no while loop inside). Thus, the overall computation complexity of finding a stable matching is $\mathcal{O}((N + M)m)$.

# 5   Performance Evaluation

## 5.1   Complexity Analysis

The primary difference between the GS algorithm and RPTS algorithm lies in
their adaptability to network dynamics. Each time, the GS algorithm starts from
an empty matching and by proposing/rejecting actions to reach a stable matching,
while the RPTS algorithm begins with the matching from the previous round and
takes the divorce/remarry operations as its path to stability. Apparently, RPTS takes
advantage of the relations between matchings in adjacent times. The computation
complexity or say iteration times for both algorithms depends on the number of
users and how fast the network changes.

As provided in Sect. 4.2, the complexity of GS is $\mathcal{O}(m)$, where $m$ is the total
length of all players' preference lists. It makes sense since the worst case of
the GS is to traverse each player's preference lists and terminate. However, the
termination condition of GS that each of the player has found its stable partner(s)
does not necessarily require the traverse of all preference lists. On the other hand,
the computation complexity of RPTS is $\mathcal{O}((N + M)m$, as indicated in Theorem 2.
Again, it is not necessary for the RPTS that all possible BPs needed to be satisfied.
Regarding the ICC algorithm, it is realized by iterative search of the currently best
BP and to swap their partners. The complexity of finding all the BPs regarding
the current matching, which requires the traverse of all users' preference lists, is
bounded by $MN$ comparing operations. On the other hand, since the swap in ICC is
irreversible, meaning each two CUs can only swap partners with each other once,
the total iterations of BP searches or swaps are bounded by $N^2$. Thus, the worst
case complexity of terminating the ICC algorithm is $\mathcal{O}(MN \times N^2)$ or simplified
as $\mathcal{O}(N^3M)$. However, the actual computation cost is not necessarily as high as
the theoretical analysis. In the simulations, we have also performed the practical
iteration times of the ICC algorithm.

Theoretically, RPTS has higher complexity than GS in the worst case, however
we should not ignore the piratical implementation. The nature of the GS structure
decides that it does not require any initial matching. However for RPTS, it can
actually take advantage of the previous matching, and transform it into stable instead
of transforming an empty matching. Thus, intuitively if some existing pairs from
the previous matching are reserved for the current period, then RPTS can save the
cost of satisfying these stable pairs. For example, if the previous matching is still
stable for the next time period, then no BP/FPBP needs to be satisfied, meaning
RPTS actually takes no action. Thus, the actual implementation complexity for
GS and RPTS may differ from the theoretical analysis. In practice, the actual
complexity depends on many complicated network factors, such as the user velocity,
network density and so on. To best evaluate the complexity for both algorithms
in wireless communication field, we quantify the complexity (convergence) by
counting the number of new connections between (CU,UU) pairs that attempted to
be set up during the whole matching procedures. However, these new connections

are not necessarily the final stable connections, since during the matching, a partnership may break up due to the deviation from any player who receives a better choice. However, building such a potential new connection requires the exchange of information through communications at both ends of the link. As we know, communication overhead is one big concern in protocol/mechanism design w.r.t. both cost and time efficiency. Thus, numerating the number of new potential link set up is in fact a reasonable measurement of the complexity cost for practical implementation. More details of both algorithms' performances are discussed in Sect. 5.2.

## 5.2   Experimental Set-Up

In this simulation, we have adopted two mobility models to test our proposed algorithms. Among many mobility models, the RWP and HotSpot models represent unpredictable and predictable user motion, respectively. The RWP model is a popular mobility model to evaluate mobile ad hoc network routing protocols due to its simplicity and wide availability. In the RWP model, the movement of nodes is governed in the following manner: each node begins by pausing for a fixed duration. Then each node selects a random destination and a random speed between 0 and the maximum velocity. The node moves to this destination and again pauses for a fixed period. This behavior is repeated until the end of simulation [14]. On the other hand, the HotSpot model is also commonly seen. For example, people go to different places for work, dining, shopping and so on, and thus hotspots are formed. More specifically, in the HotSpot model [15], users are initially placed in the neighborhood of a point, which is called the event point. In this motion, each user moves toward the closest event point, never going closer than a minimum separation distance from the event point. Then after a fixed time interval from the start of the event (i.e., the completion of the event), users return to their original locations. Users move at a random speed between 0 and the maximum velocity, which can be changed for topology analysis.

We simulate a cellular network consisting of $B_1 = 5$ eNBs randomly distributed within a circular area with radius of $R = 0.5$ km. The number of CUs and UUs, namely $N$ and $M$, are within the range from 30 to 65, and are initially randomly distributed within the network. The number of WAPs is set as and $B_2 = 20$. We assume the total number of unlicensed spectrum as $K = 20$. The $K$ unlicensed bands are randomly allocated to the $M$ UUs. The performances of the GS and RPTS algorithms are evaluated under two mobility patterns: (1) RWP model, (2) HotSpot model. We set the simulation time slot $\Delta T$ to be 10 ms, which is selected according to the time scale of the channel slow fading. Compared with the channel fading time scale, users mobility time scale are relatively large. In order to exhibit the influence of both channel change and user movement on the resource allocations, we have made some assumptions to suit the time scale of user mobility models to that of channel fading. In the RWP model, the stop time is set as 2 ms for all users. In the HotSpot model, the fixed time interval (from the start of the event to the end) as 300 ms, which is long enough to cover 15 simulation periods so that during this time interval users are either gathering or leaving the event point. We set the total simulation time for each experiment as 150 ms, i.e., 15 time slots.[2] The maximum velocities for both RWP and HotSpot model are set to 50 m/s for CUs, while for UUs, the velocity is set as 10 m/s. An illustration of the user mobility traces are shown in Fig. 3 for both RWP and HotSpot models.[3] The bandwidth of each unlicensed band is set within [2, 4] MHz. The SINR requirement for CUs is a uniform random distribution within [20, 30] dB. While the maximum interference for VUs is $-90$ dBm (the noise level of unlicensed spectrum). For the propagation gain, we set the pass loss constant $C$ as $10^{-2}$, the path loss exponent $\alpha$ as 4, the multipath fading gain as 1, and the shadowing gain as the log-normal distribution with 4 dB deviation [16]. The channel conditions are assume to change every 10 ms time slot. The fast fading is assumed to be static during each time slot.

## 5.3   *Experimental Results*

We first analyze the impact of network dynamics on the resource allocations, caused by the channel change and user mobility, in the time frame. Figures 4, 5 and 6 evaluate the time dynamic performances of GS, RPTS and ICC algorithms, w.r.t. the computation complexity, matching update ratio, and system throughput.

The complexity (measured by the number of new connections as discussed in Sect. 5.1) of the three proposed algorithms are compared under RWP and HotSpot patterns in Figs. 4a and 5a, respectively. Apparently, the RPTS algorithm achieves a much lower complexity than GS under both mobility models during the whole 150 ms simulation period, which is about only 40% complexity of GS except at the starting point. As the theoretical analysis indicate that RPTS has higher complexity than the GS algorithm, however the practical cost depends on the network implementation. Thus, it demonstrates the effectiveness of the RPTS algorithm in transforming a random matching into stable with lower complexity than the GS algorithm. For the starting point, it's reasonable that RPTS has a relatively high cost, still lower than the GS, since it starts from am empty matching. Comparing the two curves of the ICC algorithm implemented after the GS and RPTS in both Figs. 4a and 5a, they achieve similar results, and the complexity costs for both are about eight averagely. It means that using the proposed ICC procedures, only around eight actual swaps are needed to re-stabilize the whole matching. In the HotSpot model, the complexities for all three algorithms slowly decreases as time

---

[2]This time interval is much shorter than practical case of user gathering or moving randomly, which is specifically shortened to model the impact of the motion pattern on the designed protocols.

[3]To better illustrate the user mobility traces on the drawing figures, we have tuned the maximum speeds so that the mobility traces can be evident to see.

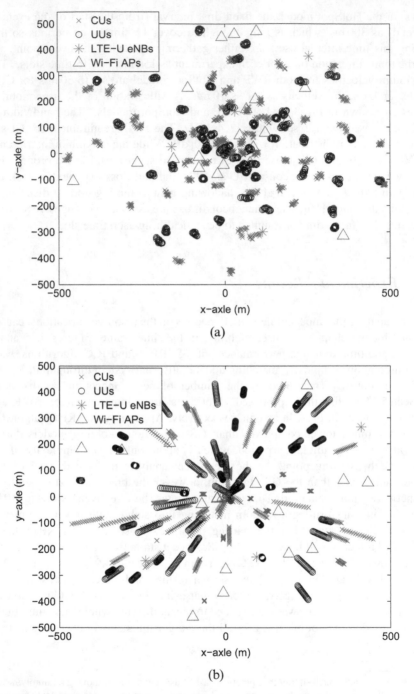

**Fig. 3** User mobility traces for both RWP mobility (**a**) and HotSpot mobility (**b**)

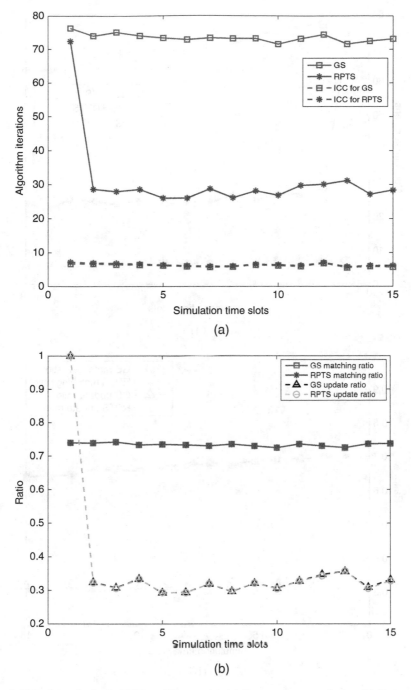

**Fig. 4** Time dynamics in the RWP mobility model. (**a**) Computation complexity. (**b**) Connection updated ratio

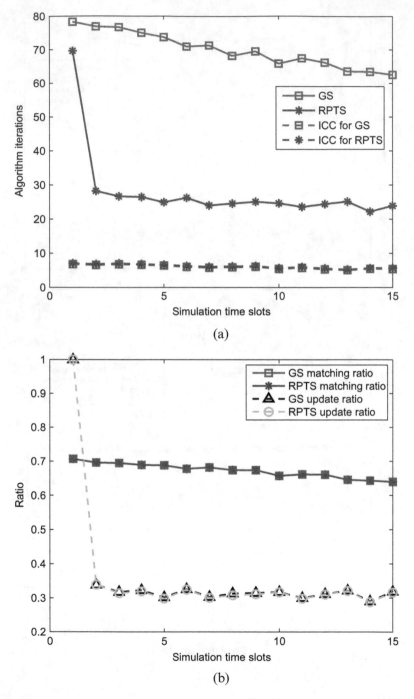

**Fig. 5** Time dynamics in the HotSpot mobility model. (**a**) Computation complexity. (**b**) Connection update ratio

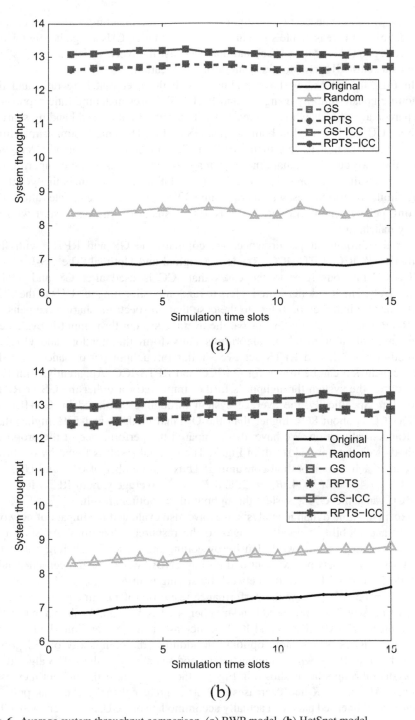

**Fig. 6** Average system throughput comparison. (**a**) RWP model. (**b**) HotSpot model

evolves, which is caused by the slight decrease of the matching ratio as indicated in Fig. 5b. This is reasonable since in the HotSpot model, CUs are gathering toward the event point (faster than UUs) and thus less distributively, which gives CUs less options as most UUs are still far from the event location.

In Figs. 4b and 5b, we have evaluated both the user matching ratio and the matching update ratio by using GS and RPTS. The user matching ratio represents the percentage of CUs who are allocated with a proper unlicensed band by sharing with an UU. As indicated in both figures, GS and RPTS achieve similar matching ratio, which is as high as 75% in the RWP and 70% in the HotSpot model, averagely. The other two curves evaluate the percentage of updated users by comparing the matching results in the previous and current simulation slots. Again, both algorithms have similar performances, which are around 30% averagely. The update ratio at the starting point for both GS and RPTS is 100%, since we assume to start with an empty matching.

For the throughput performance, we compare the GS and RPTS, with five methods: GS-ICC, RPTS-ICC, Random, Original and Optimal. The GS-ICC and RPTS-ICC methods refer to the cases that ICC is used after GS and RPTS, respectively. The Random method refers to randomly allocating the UUs to the CUs, while the Original method refers to the case that no spectrum sharing happens. In the RWP model, as shown in Fig. 6a, the average system throughput is evaluated. Apparently all four matching algorithms outperform the Random and Original methods a lot. GS and RPTS achieve similar throughput performance, and the same conclusion can be drawn for GS-ICC and RPTS-ICC. Apparently, with ICC procedures, the system throughput is further improved for either the GS or RPTS algorithm. Specifically, the average system throughput achieved by GS+ICC or RPTS+ICC is about 86% higher than the Original method, and 53% higher than the Random allocation. We have also compared the performance of the proposed methods with the optimal result in Fig. 7. The optimal result is found by the brute force approach, which is time-consuming. Thus, the number of CUs and UUs are set as $N = 4$ and $M = 4$, $B_1 = 2$, and $B_2 = 2$. Averagely, both RPTS-ICC and GS-ICC can achieve 75% system throughput of the optimal result.

Except the time dynamic analysis, we have also evaluated the impact of network density and mobility velocity changes to the resource allocations. As shown in Fig. 8, we change the network density by adding more users, including both CUs and UUs, to the network without adding any eNBs, WAPs, or unlicensed bands. We add 5 CUs and UUs to the network by staring with $N = M = 20$ and end at $N = M = 65$. We average the performance result of 150 ms time period for each network density. The unlicensed band number is set as $K = 30$. As shown in Fig. 8a, the complexity of GS, RPTS and ICC all increase as more users join the network, since more users brings more options. In addition, the complexity of GS grows faster than the RPTS, which demonstrates good scalability of the RPTS algorithm. For system throughput, as shown in Fig. 8b, the average user throughput increases before $N/M$ reaches $K$ and decreases as $N/M$ is greater than $K$. The peak point is when each unlicensed band can actually accommodate one CU, and when more CUs come after this point, the unlicensed bands will be shared between multiple CUs by TDMA.

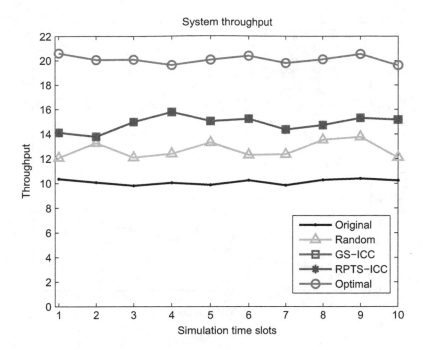

**Fig. 7** System throughput comparison with optimal solution

We changed the maximum velocity value in both mobility models to test our proposed algorithms. As shown in Fig. 9, we increase the maximum use velocity from 20 to 60 m/s by step of 5 m/s for the CUs. Apparently, the velocity changes does not necessarily has impact on the computation complexity or the system throughput, which on the other hand validate that our assumptions in the two user mobility models have no impact on the results although slightly different from the practical case.

## 6  Summary

In this chapter, we have studied the dynamic resource allocation problem in the U-LTE in a semi-distributive manner. The SM matching model well has interpreted the two-sided feature of the resource allocation system. The proposed GS and RPTS algorithms provide close optimal system performance, while both guaranteeing system QoS requirements and stability. Specifically, the RPTS algorithm, different from the repeated static resource allocation GS, achieves better performance w.r.t. the practical implementation complexity, CU matching ratio, and matching update ratio. In other words, the RPTS algorithm is more adaptable than the GS algorithm under both unpredictable and predictable mobility patterns in providing paths to dynamic stability in the U-LTE.

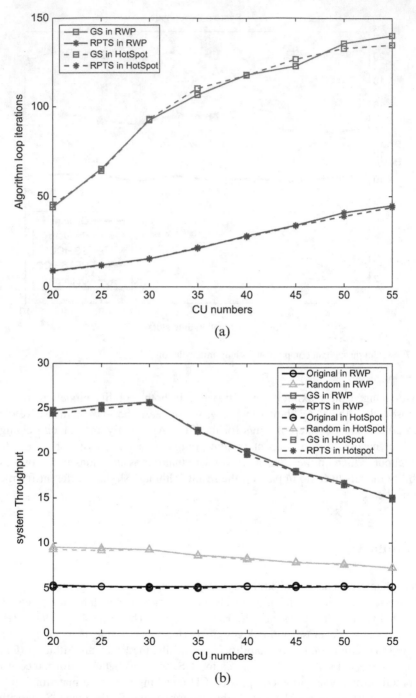

**Fig. 8** CU density dynamics in RWP&HotSpot mobility model. (**a**) Computation complexity. (**b**) System throughput

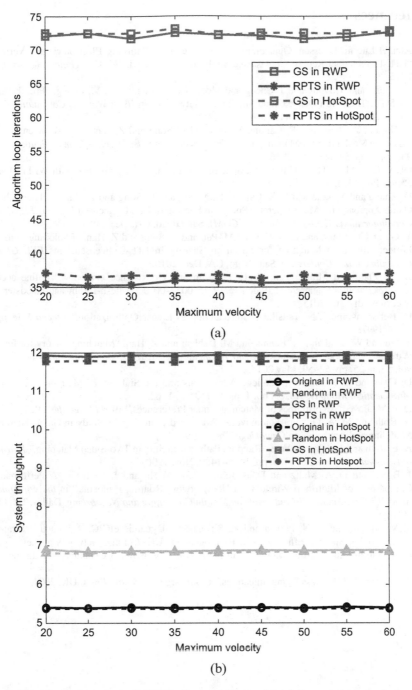

**Fig. 9** CU velocity dynamics in RWP&HotSpot mobility model. (**a**) Computation complexity. (**b**) System throughput

# References

1. Alcatel-Lucent, Ericsson, Qualcomm Technologies Inc., Samsung Electronics, and Verizon, "LTE-U Technical Report: Coexistence Study for LTE-U SDL V1.0," *Technical Report,* Feb. 2015.
2. H. Zhang and Y. Dong and J. Cheng and Md. J. Hossain and V. C. M. Leung, "Fronthauling for 5G LTE-U Ultra Dense Cloud Small Cell Networks," in *IEEE Wireless Communications,* vol. 23, no. 6, pp. 48–53, Dec. 2016.
3. Y. Gu and Y. Zhang and L. Cai and M. Pan and L. Song and Z. Han, "LTE-Unlicensed Co-existence Mechanism: A Matching Game Framework," in *IEEE Wireless Communications,* vol. 23, no. 6, pp. 54–60, Dec. 2016.
4. QUALCOMM, "LTE in Unlicensed Spectrum: Harmonious Coexistence with Wi-Fi" *White Paper,* Jun. 2014.
5. H. Zhang and Y. Xiao and L. X. Cai and D. Niyato and L. Song and Z. Han, "A Hierarchical Game Approach for Multi-Operator Spectrum Sharing in LTE Unlicensed," in *in IEEE Global Communications Conference (GLOBECOM),* San Diego, CA, Dec. 2015.
6. Y. Gu and Y. Zhang and L. X. Cai and M. Pan and L. Song and Z. Han, "Exploiting Student-Project Allocation Matching for Spectrum Sharing in LTE-Unlicensed," in *IEEE Global Communications Conference,* San Diego, CA, Dec. 2015.
7. Jingyi Zhou and Mingxi Fan, "LTE-U Forum and Coexistence Overview," online access at "https://mentor.ieee.org/802.19/dcn/15/19-15-0057-00-0000-lte-u-forum-and-coexistence-overview.pdf," Jul. 2015.
8. D. Bertsimas and J. N. Tsitsiklis, "Introduction to Linear Optimization," *Athena Scientific,* USA, 1997.
9. Y. Gu and W. Saad and M. Bennis and M. Debbah and Z. Han, "Matching Theory for Future Wireless Networks: Fundamentals and Applications," in *IEEE Communications Magazine,* vol. 53, no. 5, pp. 52–59, May 2015.
10. D. Gale and L. S. Shapley, "College Admissions and the Stability of Marriage," *American Mathematical Monthly,* vol. 69, no. 1, pp. 9–15, Jan. 1962.
11. D. F. Manlove, "Algorithmics of Matching Under Preferences," *World Scientific,* 2013.
12. A. Roth and M. A. Oliveira Sotomayor, "Two-Sided Matching: A Study in Game-Theoretic Modeling and Analysis," in *Cambridge Press,* 1992.
13. A. E. Roth and J. H. Vande Vate, "Random Paths to Stability in Two-Sided Matching," *Journals of Econometrica,* vol. 58, no. 6, pp. 1475–1480, Nov. 1990.
14. J. Broch and D. A. Maltz and D. B. Johnson and Y. Hu and J. Jetcheva, "A Performance Comparison of Multi-hop Wireless Ad Hoc Network Routing Protocols," in *the 4th Annual ACM/IEEE International Conference on Mobile Computing and Networking,* Dallas, TX, Oct. 1998.
15. S. Vasudevan and R. N. Pupala and K. Sivanesan, "Dynamic eICIC: A Proactive Strategy for Improving Spectral Efficiencies of Heterogeneous LTE Cellular Networks by Leveraging User Mobility and Traffic Dynamics," in *IEEE Trans. on Wireless Commun.,* vol. 12, no. 10, pp. 4956–4969, Oct. 2013.
16. A. Goldsmith, "Wireless Communications," *Cambridge University Press,* UK, 2004.

# Traffic Offloading from Licensed Band to Unlicensed Band

## 1 Development of Traffic Offloading

In view of the increasing requirements of mobile data rate and data applications, it becomes challenging for the traditional wireless network to meet the demands of all user equipments (UEs). Accordingly, it is beneficial and necessary to offload wireless traffic to other vacant resources. Generally, the traffic offloading is currently considered in the perspective of network architecture and wireless resource. In this section, we briefly introduce the traffic offloading in wireless communication and highlight the promising trends for traffic offloading from licensed spectrum to unlicensed spectrum with U-LTE.

### 1.1 Traffic Offloading in Heterogeneous Networks

The decoupling of the increasing density and variety of data services and the limited amount of wireless resource motivate the improvement of spectrum efficiency in wireless communication. Heterogeneous network, where multi-tier small cells overlaid on the traditional macrocells, becomes an effective solution.

In the heterogeneous network as shown in Fig. 1, the data traffic from the macrocell base station can be offloaded to the small cell base stations close to the UEs. Due to the small distance or indoor data transmission between small cell base station and UEs, low power transmission while high quality of service can be achieved, where the same spectrum can be reused multiple times between different small cell base station and UE pairs with tolerable interferences in the same macrocell. Thus, the general network efficiency is improved, leading to high quality of service (QoS) of all UEs.

© The Author(s) 2018
H. Zhang et al., *Resource Allocation in Unlicensed Long Term Evolution HetNets*,
SpringerBriefs in Electrical and Computer Engineering,
https://doi.org/10.1007/978-3-319-68312-6_5

**Fig. 1** Traffic offloading in heterogeneous network

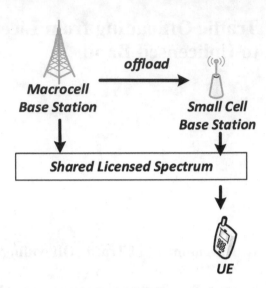

## 1.2   Traffic Offloading to Wi-Fi Networks

However, even though increasing high spectrum efficiency is able to relief the traffic congestion, with the demands of data traffic exponentially increasing, wireless operators also seek for more wireless spectrum resource to meet the requirement of all UEs. On the other hand, with the fast development of Wi-Fi technology, from 802.11 to 802.11 ac, 802.11 ad and 802.11 ax, the data transmission rate nowadays has reached up to 6.7 Gbit/s. Therefore, as shown in Fig. 2, it is promising to offload the congested data traffic in wireless communication network to the Wi-Fi networks, where the unlicensed spectrum of 2.4, 5 and 60 GHz bands in Wi-Fi is able to relief the congested traffic in wireless communication network. According to the Cisco Visual Networking Index, in 2016, 60% of total mobile data traffic was offloaded through Wi-Fi or femtocell [1]. Companies like AT&T has established and operate more than 30,000 public Wi-Fi hotspots for their wireless service offloading.

## 1.3   Traffic Offloading to Unlicensed Spectrum with U-LTE

Nevertheless, compared with Wi-Fi, the LTE technology is able to achieve higher performance for UEs. Accordingly, it is beneficial if LTE can be applied in unlicensed spectrum and the congested data traffic is able to be offloaded to the U-LTE, as shown in Fig. 3. However, as various other data services, e.g., Wi-Fi, are also presented in the unlicensed spectrum. In order to improve the Quality of Service (QoS) of its own user while guaranteeing the performance of other unlicensed users at the same time, it remains challenging for each operator to employ spectrum

**Fig. 2** Traffic offloading to
Wi-Fi Networks

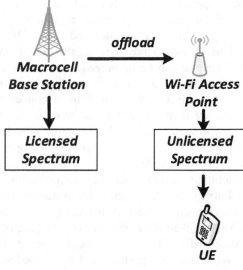

**Fig. 3** Traffic offloading to
unlicensed spectrum with
U-LTE

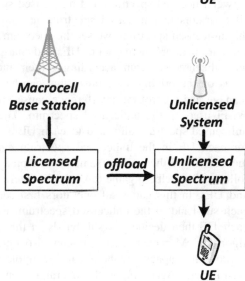

allocation between licensed spectrum and unlicensed spectrum. Moreover, when multiple operators offload their service on the unlicensed spectrum simultaneously, further power control and sub-band allocation strategies are required to avoid strong interference.

From the operators' perspective, how to manage the resource allocation in both licensed and unlicensed spectrum is a critical challenge. To minimize the interference caused by the UEs in U-LTE, a dynamic traffic balancing algorithm over licensed and unlicensed spectrum was proposed for Integrated Femto-WiFi and Dual-Band Femtocell in [2]. It is shown that the algorithm can improve the overall user experience in both licensed and unlicensed bands. In [3], a flexible resource allocation scheme is proposed to improve the efficiency of resource utilization in both licensed and unlicensed bands. By adjusting the resource on licensed and unlicensed bands dynamically based on the utility functions, the network performance can be optimized to attain the maximum utility. In [4], the authors jointly consider the power control and spectrum allocation in both licensed and unlicensed bands. With the help of convex optimization methods, the spectrum efficiency is maximized in the system. In [5], the authors propose the channel selection strategies for U-LTE enabled cells. By adopting the distributed Q-learning mechanism for channel selection, all LTE operators are able to coexist in an efficient way. In [6], a student-project allocation matching is applied to approach a stable matching results of channel allocation problem in the unlicensed spectrum.

Furthermore, continuing the system model in last chapter, the multi-operator scenarios should be considered in offload problems, where each operator tries to offload their data service from its unique but congested licensed spectrum to unlicensed spectrum. Accordingly, the resource management for each operator between licensed spectrum and unlicensed spectrum and the resource sharing for all operators in unlicensed spectrum are supposed to be jointly considered. In the unlicensed spectrum, we set the spectrum sharing scenarios in which multiple cellular operators serve a set of UEs and charge penalty prices to all UEs accessing the unlicensed spectrum according to their interference to the Wi-Fi networks. We focus on the pricing mechanism that can be applied by the cellular operators to manage and control the interference caused by each UE to other UEs as well as Wi-Fi users in the unlicensed spectrum. The amount of licensed spectrum and unlicensed spectrum allocated to each UE as well as the optimal transmit power for each UE in the unlicensed spectrum can be determined under the pricing mechanism of the operators. In this chapter, we formulate a multi-leader multi-follower Stackelberg game to study the interactions between the cellular operators and UEs. In this game, all operators first set their interference penalty price on each sub-band of the unlicensed spectrum. Based on the prices set by operators, each UE then decides its sub-bands in the unlicensed spectrum by a matching algorithm. Moreover, each UE can also optimize its transmit power to further improve its capacity without causing intolerable interference to other UEs and Wi-Fi users. Accordingly, the operators can predict the actions of the UEs and set the optimal prices to receive high utilities. We propose both non-cooperative and cooperative schemes for operators to deal with the interference problem in the unlicensed spectrum. In the non-cooperative scheme, each operator sets its prices individually without coordinating with others, and a sub-gradient algorithm is adopted to achieve the highest utility for each operator based on the behaviors of others. In the cooperative scheme, all operators are able to coordinate when they set prices. We optimize the relations of the prices with a linear programming method so as to reach the highest utilities of all operators. To the best of our knowledge, this is the first work that applies the Stackelberg game with multiple leaders and multiple followers to study the U-LTE networks. Simulation results show that the operators in both the non-cooperative and cooperative schemes can improve their utilities without causing intolerable interferences to the unlicensed users, based on different traffic conditions in the unlicensed spectrum.

The rest of this chapter is organized as follows. We introduce the system model in Sect. 2, and then formulate the problems in Sect. 3. Based on the formulated problem, we model the scenario in a multi-leader multi-follower Stackelberg game and further analyze the game in Sects. 4 and 5. We present our simulation results in Sect. 6 and finally summarize this chapter in Sect. 7.

# 2 System Model

We consider a heterogenous cellular network system where $M$ co-located operators serve $N$ UEs in an indoor environment. We assume operator $i$, $\forall i \in \mathcal{M} = \{1, 2, \ldots, M\}$, deploys $P_i$ Small Cell Base Stations (SCBSs) that are co-located with $Q_i$ Wi-Fi Access Points (WAPs), randomly distributed in the coverage area. The SCBSs can serve the UEs in both the licensed and unlicensed spectrum. In the licensed spectrum, we assume all UEs operate in the same manner as the traditional LTE networks and are able to obtain licensed resource that can support $C_j^l$ data transmission rate, $\forall j \in \mathcal{N} = \{1, 2, \ldots, N\}$. If UE $j$ is satisfied with a data transmission rate that is less than or equal to $C_j^l$, it will only access the licensed spectrum. If UE $j$ requires a data transmission rate that is higher than $C_j^l$, UE $j$ will then also seek spectrum resource in the unlicensed spectrum to further improve its Quality-of-Service (QoS). To simplify our description, we assume the channel gains between cellular base station and UEs can be regarded as constants, and therefore $C_j^l$ can be regarded as a fixed value so that we can focus on the resource allocation in the unlicensed spectrum. In each sub-band of both licensed and unlicensed spectrum, we suppose there is an upper bound on the transmit power. As the resource management mechanisms in the licensed spectrum are currently mature and well-deployed in the telecommunication network, in order to adopt U-LTE without affecting the original resource management, we follow the current power control mechanism in the licensed spectrum first. If the UEs are not satisfied with the services in licensed spectrum, following the power constraint in each sub-band, the power control in the unlicensed spectrum is executed. Suppose $N$ UEs require to access to the unlicensed spectrum. In the unlicensed spectrum, all operators utilize a common spectrum pool with Wi-Fi access points and other unlicensed users. In order to guarantee the performance of other unlicensed users, the transmit power of each UE cannot strongly interfere with other unlicensed users in the same sub-band, or surpass the available residue power. Furthermore, we assume that the UEs served by the SCBSs can be allocated with unlicensed spectrum, and that each UE chooses the operator with the SCBS closest to it. We suppose there are $S$ sub-bands in the unlicensed spectrum. When multiple UEs are allocated with the same sub-band in the unlicensed spectrum, the UEs may cause severe interference among each other. Accordingly, we follow the same setting as our previous works [7] and consider the dynamic spectrum access systems with multiple operators. We assume all the operators can share the unlicensed spectrum with Wi-Fi networks. Each operator

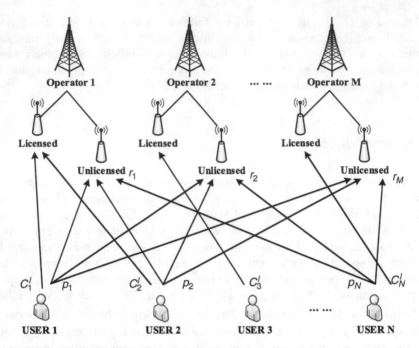

**Fig. 4** System architecture in multi-operator multi-user scenario

can access any sub-band that is occupied or unoccupied by Wi-Fi users in the spectrum pool. However, each sub-band can only be accessed by one operator at each time. For the UEs served by the same operator in U-LTE, the LTE standard is applied in the unlicensed spectrum. Thus, Orthogonal Frequency Division Multiple Access (OFDMA) is adopted to avoid cross-interference. For UEs that are served by different operators, we suppose that Frequency Division Multiple Access (FDMA) is applied [8]. As shown in Fig. 4, in the unlicensed spectrum, following the settings in [9, 10, 12], before the data transmission between each UE and its serving SCBS, in the control channels, the operators are able to broadcast the prices that it would charge in the unlicensed spectrum to all the UEs because of the interference to the Wi-Fi users. Based on the prices set by all the operators, UE $j$, where $j \in \mathcal{N}$, determines its desired transmit power in the sub-band $s$, $\forall s \in \mathcal{S} = \{1, 2, \dots, S\}$, which is denoted as $p_{j,s}$.

When UE $j$ is served by the operator $i$ in the sub-band $s$, $\forall s \in \mathcal{S}$, of the unlicensed spectrum, we define the spectrum efficiency of UE $j$ as

$$R_{j,s} = \log_2 \left( 1 + \frac{p_{j,s} g_j}{Z_{j,s}} \right), \tag{1}$$

where $g_j$ is the channel gain from the serving SCBS to UE $j$, $Z_{j,s}$ is the total interference measured by UE $j$ in the sub-band $s$. Receiving the training data, the serving SCBS are able to feedback the estimated channel response $g_j$ and interference $Z_{j,s}$ to UEs for decisions [13].

Accordingly, we suppose $B_u$ is the size of each sub-band in the unlicensed spectrum. If UE $j$, $\forall j \in \mathcal{N}$, is served in both the licensed and unlicensed spectrum, the utility of UE $j$ can be shown as

$$U_j = C_j^l + \sum_{s=1}^{S} \lambda_{j,s} \left( \gamma_j B_u R_{j,s} - \sum_{i=1}^{M} \sum_{k=1}^{Q_i} r_i h_{ikj} p_{j,s} \right), \tag{2}$$

where $\gamma_j B_u R_{j,s}$ is the profit that UE $j$ receives from the services in the sub-band $s$, $\forall s \in \mathcal{S}$, of the unlicensed spectrum. $\gamma_j$ is the revenue that UE $j$ gains for unit data rate transmitted. $r_i$ is the penalty price for unit watt of operator $i$ in the unlicensed spectrum, $h_{ikj}$ is the channel gain from the $k$th WAP of operator $i$ to UE $j$, and $p_{j,s}$ is the transmit power of UE $j$ in the sub-band $s$, $\forall s \in \mathcal{S}$, of the unlicensed spectrum. As the data transmission in the unlicensed spectrum causes interference to the WAPs nearby, we set $r_i p_{j,s} h_{ikj}$ as the interference penalty from the $k$th WAP of operator $i$ to UE $j$ in the sub-band $s$ of the unlicensed spectrum, $k \in \mathcal{K}_i = \{1, 2, \ldots, Q_i\}$, $i \in \mathcal{M}$, $\forall s \in \mathcal{S}$. The WAPs of operators can forward the information to the core communication network and feedback the estimated channel gain $h_{ikj}$ to UEs for decisions. $\lambda_{j,s}$ is a binary number determining whether or not the sub-band $s$ is allocated to UE $j$.

Accordingly, the utility of operator $i$ is defined as the revenues received from all WAPs of the operator to all the UEs in the unlicensed spectrum, i.e., $\forall i \in \mathcal{M}$,

$$W_i = r_i \sum_{s=1}^{S} \sum_{j=1}^{N} \left( \lambda_{j,s} p_{j,s} \sum_{k=1}^{Q_i} h_{ikj} \right). \tag{3}$$

## 3 Problem Formulation

In a cellular network system with multiple operators and UEs, it is possible that not every operator is always interested to coordinate with others. We therefore consider two specific scenarios: all the operators can either non-cooperate with each other or can fully coordinate with each other by forming as a group. When some operators cooperate and do not cooperate, we can combine the above two situations and solve the problem.

When the operators are not fully coordinated with each other, they can make decisions in a distributed manner, i.e., operator $i$ sets its price $\mathbf{r}_i$ of the interference penalty to all UEs served on all sub-bands in the unlicensed spectrum. Not only should it predict the reactions of all the UEs, but it also needs to consider the behaviors of other operators in order to receive satisfying revenues. Therefore, the optimization problem for operator $i$ is,

$$\max_{r_i} \quad W_i(r_i \mid \mathbf{r}^*_{-i}, \mathbf{p}^*), \qquad \forall i \in \mathcal{M},$$

$$s.t. \begin{cases} \mathbf{r}^* \geq \mathbf{0}, \\ p^*_{j,s} \geq 0, & \forall j \in \mathcal{N}, \ \forall s \in \mathcal{S}, \\ p^*_{j,s} < p^{\max}_{j,s}, & \forall j \in \mathcal{N}, \ \forall s \in \mathcal{S}, \end{cases} \tag{4}$$

where $\mathbf{r}^*_{-i}$ is the set of the optimal pricing strategies of all other operators except operator $i$ on all sub-bands of the unlicensed spectrum. $\mathbf{r}^* = [r^*_1, r^*_2, \ldots, r^*_M]$ is the set of the optimal pricing strategies of all operators. $\mathbf{0} = [0, 0, \ldots, 0]$ is the set with M elements, each of which is zero. $\mathbf{p}^* = [p^*_1, p^*_2, \ldots, p^*_N]$ is the set of the optimal transmit powers of all UEs on all sub-bands of the unlicensed spectrum. In order to manage the interference to ensure the service of unlicensed users nearby, the operators should control the transmit power of each UE. We define $p^{\max}_{j,s}$ as the maximum transmit power of UE $j$ in the sub-band $s$ of the unlicensed spectrum, $\forall j \in \mathcal{N}, \ \forall s \in \mathcal{S}$.

Furthermore, when all operators are able to cooperate with each other, all operators aim to achieve the maximum total utility. Accordingly, before setting prices of interference for all UEs in the unlicensed spectrum, operators are only required to predict the transmit power of all UEs so as to achieve high utilities. The optimization problem for all operators is then formulated as follows,

$$\max_{\mathbf{r}} \quad \sum_{i=1}^{M} \alpha_i W_i(\mathbf{r}),$$

$$s.t. \begin{cases} \mathbf{r} \geq \mathbf{0}, \\ p^*_{j,s} \geq 0, & \forall j \in \mathcal{N}, \ \forall s \in \mathcal{S}, \\ p^*_{j,s} < p^{\max}_{j,s}, & \forall j \in \mathcal{N}, \ \forall s \in \mathcal{S}, \end{cases} \tag{5}$$

where $\alpha_i, \ \forall i \in \mathcal{M}$ is the weight factors for operator $i$. If $\alpha_i$ increases, operator $i$ plays a more significant role in the cooperation.

According to the optimal prices set by all operators $\mathbf{r}^*$, UE $j$ determines the transmit power strategy in each sub-band of the unlicensed spectrum $p_{j,s}$. Accordingly, the optimization problem for UE $j$ satisfies,

$$\max_{p_{j,s}, \boldsymbol{\lambda}_j} \quad U_j(p_{j,s} \mid \mathbf{r}^*, \boldsymbol{\lambda}_{-j}), \qquad \forall j \in \mathcal{N}, \ \forall s \in \mathcal{S},$$

$$s.t. \begin{cases} p_{j,s} > 0, \\ p_{j,s} < p^{\max}_{j,s}, \\ \lambda_{j,s} B_u R_{j,s} \geq \lambda_{j,s} \sum_{i=1}^{M} \sum_{k=1}^{Q_i} r_i h_{ikj} p_{j,s}, \end{cases} \tag{6}$$

where $\boldsymbol{\lambda}_j = [\lambda_{j,1}, \ldots, \lambda_{j,S}]$ is the sub-band allocation result for UE $j$, $\boldsymbol{\lambda}_{-j}$ is the sub-band allocation results for all other UEs except UE $j$. The received revenue of UE $j$, i.e., $B_u R_j$, in the serving sub-band should be no less than the interference penalty the

UE pays to all operators $\sum_{i=1}^{M} \sum_{k=1}^{Q_i} r_i h_{ikj} p_{j,s}$. As the UEs are unable to acknowledge the information of Wi-Fi users, we let the operators to set prices to restrict the transmit power of UEs. When the price imposed by each operator is high, no UE can afford the prices and therefore no UE will access the service provided by each operator. Therefore, in the formulated problem of operators, we set the power constraint for all UEs to guarantee the basic data transmission of Wi-Fi users.

Based on the above formulations, all operators and UEs are autonomous decision makers who would like to maximize their own utilities in a selfish manner. In order to analyze the problem of resource allocations in the unlicensed spectrum, we model the scenario as a multi-leader multi-follower Stackelberg game, where all operators are leaders and all UEs are followers. In the game, each operator first sets its penalty price of interference in the unlicensed spectrum. Based on the prices set by all operators, each UE determines its optimal transmit power. In the following sections, backward induction is adopted to analyze the problems. We first discuss the strategy of each UE, given the penalty price of interference set by all operators. Then, with the prediction of the optimal behaviors of each UE, we design a sub-band allocation scheme with matching theory and propose the corresponding non-cooperative or cooperative strategies for operators to achieve the maximum utilities.

# 4 Analysis of UEs

Observing the prices set by operators, the UEs are supposed to adopt strategies for optimal utilities. In this section, we first analyze the optimal power transmission strategies for the UEs. Based on the optimal transmit power on each sub-bands of the unlicensed spectrum, we then design a sub-band allocation scheme with matching theory for high utilities.

## 4.1 Strategies of Power Transmission for UEs

In the formulated multi-leader multi-follower Stackelberg game, all UEs act as followers. In order to receive high revenues from the services and reduce the interference penalty to other operators, based on the prices set by operators $i$, $\forall i \in \mathcal{M}$, UE $j$ optimizes its transmit power $p_{j,s}$ in the sub-band $s$ of the unlicensed spectrum, $\forall j \in \mathcal{N}$, $\forall s \in \mathcal{S}$. The optimal transmit power for each UE is relative to the prices set by all operators. Lemma 1 is developed as follows.

**Lemma 1** *If UE $j$ is served by operator $i$ in the unlicensed spectrum, $\forall i \in \mathcal{M}$, $\forall j \in \mathcal{N}$, the optimal transmit power to UE $j$ on the sub-band is*

$$p_{j,s}{}^* = \left( \frac{B_u}{\sum\limits_{i=1}^{M} \sum\limits_{k=1}^{Q_i} h_{ikj} r_i} - \frac{1}{q_{j,s}} \right)^+, \tag{7}$$

*where*

$$(x)^+ = \max \{x, 0\}, \tag{8}$$

*and*

$$q_{j,s} = \frac{g_j}{Z_{j,s}}. \tag{9}$$

In (7), as the channel gain $g_{j,s}$ is related to the distance between UE $j$ and its serving SCBS, and the channel gain $h_{ikj}$ is related to the distance between the $k$th WAP of the operator $i$ and UE $j$, we discover that when the distance between UE $j$ and its serving SCBS increases, the channel gain $g_{j,s}$ decreases. Thus, the optimal transmit power $p_{j,s}$ in the sub-band $s$ decreases. When the distances between the UE $j$ and the $k$th WAP of the operator $i$ increases, the value of channel gain $h_{ikj}$ decreases. Thus the optimal transmit power $p_{j,s}$ in the sub-band $s$ increases.

*Proof* When UE $j$ is allocated with the unlicensed spectrum, the utility function of UE $j$ is continuous. We take the second derivative of $U_j$ with respect to $p_{j,s}$, i.e., $\forall s \in \mathcal{S}$,

$$\frac{\partial^2 U_j}{\partial p_{j,s}{}^2} = -\frac{B_u q_{j,s}^2}{(1 + p_{j,s} q_{j,s})^2}. \tag{10}$$

The second derivative of $U_j$ with respect to $p_{j,s}$ is negative, so $U_j$ is quasi-concave in $p_{j,s}$. Accordingly, when the first derivative of $U_j$ with respect to $p_{j,s}$ is equal to zero, i.e., $\forall s \in \mathcal{S}$,

$$\frac{\partial U_j}{\partial p_{j,s}} = \frac{B_u q_{j,s}}{1 + p_{j,s} q_{j,s}} - \sum_{i=1}^{M} \sum_{k=1}^{Q_i} h_{ikj} r_i = 0, \tag{11}$$

the utility function of UE $j$ achieves the maximum value, where the transmit power from the operator $i$ to UE $j$ in the sub-band $s$, $\forall s \in \mathcal{S}$, of the unlicensed spectrum satisfies

$$p_{j,s} = \frac{B_u}{\sum\limits_{i=1}^{M} \sum\limits_{k=1}^{Q_i} h_{ikj} r_i} - \frac{1}{q_{j,s}}. \tag{12}$$

Furthermore, the transmit power $p_{j,s}$ follows the constraint $p_{j,s} \in [0, p_{j,s}^{\max}]$. On one hand, according to the properties of quasi-concave function, if the value of (12) is negative, the optimal solution in the feasible region is $p_{j,s}^{*} = 0$, i.e., there are many other UEs and unlicensed users transmitting information on the sub-band $s$ of the unlicensed spectrum. Thus, the transmit power on the sub-band is zero because of the high interference penalty. On the other hand, each UE is unaware of the interference it will cause to other unlicensed users when it accesses each sub-band. For UE $j$, if $p_{j,s}$ is larger than the maximum transmit power constraint $p_{j,s}^{\max}$ in the sub-band $s$ of the unlicensed spectrum, the UE $j$ will cause severe interference to all other unlicensed users in the sub-band. In order to ensure the performance of other unlicensed users, we suppose the transmit power for each UE in the unlicensed spectrum can be predicted and controlled by the operators, which will be illustrated in the following sections.

Correspondingly, when UE $j$ is served in the sub-band $s$, $\forall s \in S$, of the unlicensed spectrum, the maximum utility of UE $j$ in the sub-band, if $p_{j,s}^{*} = 0$, is

$$u_{j,s} = 0, \tag{13}$$

where $u_{j,s}$ is the utility of UE $j$ in the sub-band $s$ of the unlicensed spectrum, $\forall j \in \mathcal{N}$, $\forall s \in S$. If $p_{j,s}^{*} > 0$, we have

$$u_{j,s} = B_u \log_2 \left( \frac{q_{j,s}}{\sum\limits_{i=1}^{M} \sum\limits_{k=1}^{Q_i} h_{ikj} r_i} \right) - B_u + \frac{\sum\limits_{i=1}^{M} \sum\limits_{k=1}^{Q_i} h_{ikj} r_i}{q_{j,s}}, \tag{14}$$

where the optimal utility is related to the prices of operator $i$ in the game, $\forall i \in \mathcal{M}$. In (14), we take the second derivative of $u_{j,s}$ with respect to $r_i$, i.e.,

$$\frac{\partial^2 u_{j,s}}{\partial r_i^2} = \frac{B_u \left( \sum\limits_{k=1}^{Q_i} h_{ikj} \right)^2}{\left( \sum\limits_{i=1}^{M} \sum\limits_{k=1}^{Q_i} h_{ikj} r_i \right)^2}. \tag{15}$$

We discover $\frac{\partial^2 u_{j,s}}{\partial r_i^2} \leq 0$, i.e., the optimal utility of each UE is quasi-convex with respect to the penalty prices set by operator $i$, if the penalty prices of all other operators keep unchanged. Accordingly, we set the first derivative of $u_{j,s}$ with respect to $r_i$ equal to zero,

$$\frac{\partial u_{j,s}}{\partial r_i} = -\frac{B_u \sum\limits_{k=1}^{Q_i} h_{ikj}}{\sum\limits_{i=1}^{M} \sum\limits_{k=1}^{Q_i} h_{ikj} r_i} + \frac{\sum\limits_{k=1}^{Q_i} h_{ikj}}{q_{j,s}}. \tag{16}$$

Thus,

$$\sum_{i=1}^{M} \sum_{k=1}^{Q_i} h_{ikj} r_i = B_u q_{j,s}. \tag{17}$$

Based on the above, when the price of operator $i$ increases and the prices of all other operators are unchanged, the utility of UE $j$ first decreases. When the increasing price satisfies (17), the utility of UE $j$ stops decreasing and starts to increase as the price continuously increases.

## 4.2 Sub-band Allocation Scheme

During service, as each UE prefers to be allocated with the sub-band for high utility, we construct a preference list for UE $j$ based on the utility $u_{j,s}$ in each sub-band $s$, such that

$$PL_{UE}(j, s) = u_{j,s}. \tag{18}$$

Considering the optimal transmit power strategies of all UEs, we take the second derivative of $u_{j,s}$ with respect to $Z_{j,s}$, i.e.,

$$\frac{\partial^2 u_{j,s}}{\partial Z_{j,s}^2} = \frac{B_u}{\left(Z_{j,s}\right)^2}, \tag{19}$$

which is larger than zero, i.e., the $u_{j,s}$ is a quasi-convex function with respect to $Z_{j,s}$. Accordingly, we set the first derivative of $u_{j,s}$ with respect to $Z_{j,s}$ equal to zero, such that,

$$\frac{\partial u_{j,s}}{\partial Z_{j,s}} = -\frac{B_u}{Z_{j,s}} + \frac{\sum_{i=1}^{M} \sum_{k=1}^{Q_i} h_{ikj} r_i}{g_j} = 0. \tag{20}$$

Thus

$$Z_{j,s}^* = \frac{B_u g_j}{\sum_{i=1}^{M} \sum_{k=1}^{Q_i} h_{ikj} r_i}. \tag{21}$$

When $Z_{j,s}$ is less than $Z_{j,s}^*$, and $Z_{j,s}$ is increasing, the utility $u_{j,s}$ decreases. When $Z_{j,s}$ surpasses $Z_{j,s}^*$, the utility $u_{j,s}$ starts increasing. Moreover, according to the constraint $p_{j,s} > 0$, we have

$$Z_{j,s}^* < \frac{B_u g_j}{\sum_{i=1}^{M} \sum_{k=1}^{Q_i} h_{ikj} r_i}. \tag{22}$$

Therefore, with $Z_{j,s}$ increasing, the utility $u_{j,s}$ monotonously decreases in the available region. Accordingly, UE $j$ prefers to be served in the sub-band $s$ with low interference from other unlicensed users $Z_{j,s}$.

Moreover, we construct a preference list for sub-band $s$ based on the total revenue the operators receive from the sub-band $s$, which is denoted as $w_s$, $\forall s \in \mathcal{S}$,

$$PL_{SB}(s,j) = w_s. \tag{23}$$

Based on the predictions of all UEs' optimal strategies, the $w_s$ can be expressed as follows,

$$w_s = \sum_{i=1}^{N} \sum_{j=1}^{N} \sum_{k=1}^{Q_i} r_i \lambda_{j,s} h_{ikj} \left( \frac{B_u}{\sum\limits_{l=1}^{M} \sum\limits_{k=1}^{Q_i} h_{lkj} r_l} - \frac{Z_{j,s}}{g_j} \right). \tag{24}$$

We take the first derivative of $w_s$ with respective to $Z_{j,s}$ and discover that the value of $w_s$ is monotonously decreasing when $Z_{j,s}$ increases. Therefore, each sub-band $s$ prefers to be allocated to the UE with small interference.

Based on the preference lists from both UEs and sub-bands, we design a resident-oriented Gale-Shapley (RGS) algorithm [14] for sub-band allocation, which is shown in Algorithm 1. In Algorithm 1, each UE first proposes to its desired sub-bands based on its preference list. According to the proposal from all UEs, if more than one UE chooses the same sub-band, the sub-band keeps the most preferred UE based on its preference list and reject all the rest. The rejected UEs then continue to propose to its preferred sub-bands based on the rest of its preference list. The circulation continues until each UE is either allocated with sub-bands in the unlicensed spectrum, or rejected by all the sub-bands on their preference lists. The UE which is rejected by all the sub-bands on their preference lists will be only allocated with licensed spectrum for services.

**Lemma 2** *Following Algorithm 1, the RGS algorithm will ultimately converge and achieve a stable matching result.*

*Proof* The detailed proof can be found in [14, 15].

## 5 Analysis of Operators

Based on the predictions of the UEs' behaviors and the sub-band allocation results, we first consider that all operators are non-cooperative with each other. Each operator is required to consider the behaviors of other operators and determine its optimal strategy. Afterwards, we propose a cooperative scheme where all operators make decisions in a coordinated way so as to achieve high utility of all operators.

---

**Algorithm 1** RGS algorithm for sub-band allocation

---

1: **for** UE $j$ **do**
2:     Construct the preference list of sub-bands $PL_{UE}$ based on the value of $Z_{j,s}$;
3: **end for**
4: **for** Sub-band $s$ **do**
5:     Construct the preference list of UEs $PL_{SB}$ based on the value of $Z_{j,s}$;
6: **end for**
7: **while** the system is unmatched **do**
8:     UEs propose to sub-bands;
9:     **for** Unmatched UE $j$ **do**
10:       Propose to first sub-band $c_j$ in its preference list;
11:       Remove $c_j$ from the preference list;
12:     **end for**
13:     Sub-bands make decisions;
14:     **for** Sub-band $s$ **do**
15:       **if** 1 or more than 1 UE propose to the sub-band **then**
16:         The sub-band $s$ chooses the most preferred UE and rejects the rest;
17:       **end if**
18:     **end for**
19: **end while**

---

## 5.1 Noncooperative Strategies for Operators

In the unlicensed spectrum, based on the predictions of all UEs' optimal strategies, the utility function of operator $i$, $\forall i \in \mathcal{M}$, satisfies

$$W_i = \sum_{s=1}^{S} \sum_{j=1}^{N} \sum_{k=1}^{Q_i} \lambda_{j,s} r_i h_{ikj} \left( \frac{B_u}{\sum_{l=1}^{M} \sum_{k=1}^{Q_i} h_{lkj} r_l} - \frac{1}{q_{j,s}} \right). \tag{25}$$

Accordingly, each operator is required to determine its prices on the unlicensed spectrum for satisfactory utilities. We take the second derivative of operator $i$'s utility function,

$$\frac{\partial^2 W_i}{\partial r_i^2} = - \sum_{s=1}^{S} \sum_{j=1}^{N} \sum_{k=1}^{Q_i} 2\lambda_{j,s} b_j h_{ikj} A_j < 0. \tag{26}$$

where

$$A_j = \frac{B_u \sum_{k=1}^{Q_i} h_{ikj} \sum_{l=1, l\neq i}^{M} \sum_{k=1}^{Q_i} h_{lkj} r_l}{\left( \sum_{l=1}^{M} \sum_{k=1}^{Q_i} h_{lkj} r_l \right)^3}. \tag{27}$$

As the second derivative of $W_i$ with respective to $r_i$ is negative, $W_i$ is a concave function.

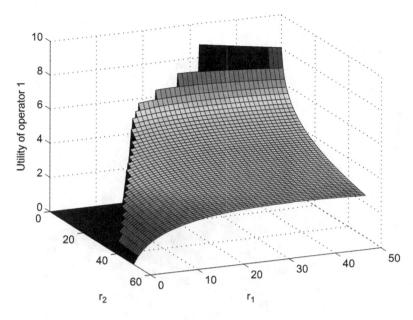

**Fig. 5** The utility of operator 1 vs. the prices set by all operators

To better analyze the problem, without loss of generality, assume there are two operators and two UEs in the unlicensed spectrum. With different prices set by both operators, the utilities of both operators are shown in Figs. 5 and 6, respectively.

In both figures, the $x$ axis denotes the price set by operator 1, and the $y$ axis is the price set by operator 2. In Fig. 5, $z$ axis refers to the utility of operator 1. In Fig. 6, $z$ axis refers to the utility of operator 2. We observe that when the prices of one operator is fixed, the utility of the other operator is a concave function of its price.

Moreover, the transmit power is constrained with $p_{j,s} \in [0, p_{j,s}^{\max}]$, $\forall j \in \mathcal{N}$, $\forall s \in \mathcal{S}$. Thus, on one hand, if the prices are set too high, no UE can afford the high payment. The optimal transmit power of each UE calculated from (7) is $p_{j,s} = 0$. In this case, operators cannot get any revenue. On the other hand, if the prices are set too low, in order to avoid interference with Wi-Fi users, the highest transmit power cannot surpass $p_{j,s}^{\max}$, resulting in low revenue for each operator. Accordingly, the price of each operator has upper and lower bounds, satisfying,

$$p_{j,s} = \frac{B_u}{\sum\limits_{i=1}^{M}\sum\limits_{k=1}^{Q_i} h_{ikj}r_i} - \frac{1}{q_{j,s}} \in \left[0, p_{j,s}^{\max}\right], \quad \forall j \in \mathcal{N}, \forall s \in \mathcal{S}. \tag{28}$$

Hence, we consider a linear combination of prices set by all operators as

$$R = \sum_{i=1}^{M}\sum_{k=1}^{Q_i} h_{ikj}r_i. \tag{29}$$

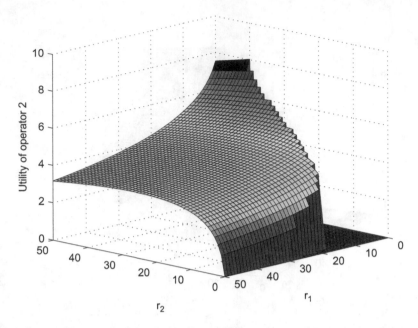

**Fig. 6** The utility of operator 2 vs. the prices set by all operators

Based on the constraints of all UEs' transmit power, for operator $i$, $\forall i \in \mathcal{M}$, the prediction of prices set by all other operators in the sub-band $s$ of the unlicensed spectrum follows the constraint,

$$R \in \left[ \frac{B_u q_{j,s}}{p_{j,s}^{\max} q_{j,s} + 1}, B_u q_{j,s} \right]. \tag{30}$$

Therefore, in order to achieve a Nash Equilibrium solution of the problem, based on the sub-band allocation results, we adopt the sub-gradient method for the pricing strategies of operators. The method is shown in Algorithm 2. In Algorithm 2, all operators start with a high price, such that no UEs would like to be served in the unlicensed spectrum. Then in each round of the circulation, for operator $i$, $\forall i \in \mathcal{M}$, we set a small step $\Delta$ and changes its current prices $r_i$ with $\Delta$ higher or lower than the original price. If the utility is the highest when the price increases with $\Delta$, in the next round, the price changes to be $r_i + \Delta$. If the utility is the highest when the price decreases with $\Delta$, in the next round, the price changes to be $r_i - \Delta$. Otherwise, the price remains unchanged. the circulation continues until all operators can not deviate from their current price unilaterally for higher utilities.

**Lemma 3** *When the starting price and the original step size $\Delta$ are fixed, the game can always converge to a unique outcome, which is also the Nash equilibrium of the game.*

---

**Algorithm 2** Strategy of operators in U-LTE

---

1: Initially, each operator sets high price. Thus, the transmits power of all UEs equal 0.
2: **while** At least one operator adjusts its price **do**
3:     **for** UE $j$ **do**
4:         Based on the price set by all operators and the sub-band allocation results, each UE determines the optimal transmit power in unlicensed spectrum.
5:     **end for**
6:     **for** operator $i$ **do**
7:         Each operator stores the current value of the service prices, $\mathbf{r}_{old} = \mathbf{r}$.
8:         Each operator tries to increase and decrease its price with a small step $\Delta = \Delta \times 0.99$, and calculates its own payoff based on the prediction of the followers' optimal strategies.
9:         **if** $R(\mathbf{r}_{old} - \Delta) < \frac{B_u q_j}{p_j^{\max} q_j + 1}$ **then**
10:            The Wi-Fi users is interfered. $W_i = -\inf$.
11:         **end if**
12:         **if** $W_i(r_{old_i}, \mathbf{r}_{old-i}) \leq W_i(r_{old_i} + \Delta, \mathbf{r}_{old-i})$ and $W_i(r_{old_i} - \Delta, \mathbf{r}_{old-i}) \leq W_i(r_{old_i} + \Delta, \mathbf{r}_{old-i})$ **then**
13:            $r_i = \min\{r_i^{\max}, r_{old_i} + \Delta\}$; % Increase the price
14:         **else**
15:            **if** $U_i(r_{old_i}, \mathbf{r}_{old-i}) \leq U_i(r_{old_i} - \Delta, \mathbf{r}_{old-i})$ and $W_i(r_{old_i} + \Delta, \mathbf{r}_{old-i}) \leq W_i(r_{old_i} - \Delta, \mathbf{r}_{old-i})$ **then**
16:                $r_i = \max\{0, r_{old_i} - \Delta\}$; % Reduce the price
17:            **else**
18:                $r_i = r_{old_i}$; % Keep the price unchanged
19:            **end if**
20:         **end if**
21:     **end for**
22: **end while**

---

*Proof* The convergence of the sub-gradient algorithm has been proved in [16] and [17]. According to [16] and [17], the sub-gradient algorithm is able to achieve an optimal solution with small ranges in convex optimization. Therefore, with given moving step size, each operator is unable to unilaterally adjust its price in order to receive higher utility when the sub-gradient algorithm converges to an optimal solution.

Furthermore, when the starting price and the original $\Delta$ are fixed, the results in the second iteration are fixed. According to the mathematical induction, we suppose that at the $Q$th iteration, the prices of operators are fixed. Then in the $(Q + 1)$th iteration, in accordance with the proposed sub-gradient strategy, the step size is fixed, and the direction from the current iteration to the next iteration is unique. Therefore, the prices of operators in the $(Q + 1)$th iteration are also fixed. Based on the above, the game can converge to a unique outcome, when the starting price and the original $\Delta$ are fixed.

## 5.2    Cooperative Strategies for Operators

Nevertheless, in order to make full use of wireless resources and achieve high revenues, some wireless operators may cooperate with each other in the unlicensed spectrum. In this subsection, we analyze the behaviors of operators when they cooperate and optimize the weighted utilities of all operators, such that,

$$W^{all} = \sum_{i=1}^{M} \alpha_i W_i. \tag{31}$$

According to the strategies of all UEs, when all operators set different prices for interference, the transmit power of UEs may be different. However, in order to avoid the interference to nearby unlicensed users, the transmit power of each UE $i$th is constrained as $p_{j,s} \in \left[0, p_j^{\max}\right]$. Therefore, if the transmit power of all UEs is maintained in a feasible region, the prices of all operators $\mathbf{r} = [r_1, r_2, \ldots, r_M]$ should satisfy

$$\sum_{i=1}^{M} \sum_{k=1}^{Q_i} h_{ikj} r_i \leq B_u q_{j,s}, \quad \forall j \in \mathcal{N}, \ \forall s \in \mathcal{S}, \tag{32}$$

$$\sum_{i=1}^{M} \sum_{k=1}^{Q_i} h_{ikj} r_i \geq \frac{B_u q_{j,s}}{p_{j,s}^{\max} q_{j,s} + 1}, \quad \forall j \in \mathcal{N}, \ \forall s \in \mathcal{S}. \tag{33}$$

Take an example of two operators in the game. We suppose there are two sub-bands in the unlicensed spectrum, which are allocated to two UEs. Following the modeling in [11], we denote the relations of pricing between operator 1 and operator 2 in Fig. 7. The $x$ axis shows the prices set by operator 1, $r_1$, and the $y$ axis shows the price set by operator 2, $r_2$. Correspondingly, according to (32), the upper bound of prices for UEs 1 and 2 are line segments AB and CD, respectively. The lower bound of prices for UEs 1 and 2 are line segments EF and GH, respectively. When both operators set prices higher than the upper bound, the UE cannot afford the interference penalty and the transmit power is zero. Therefore, in the region above CJ and JB, there are no UE served in the unlicensed spectrum. In the region BDJ, only UE 1 is served in the unlicensed spectrum. In the region ACJ, only UE 2 is served in the unlicensed spectrum. In the region AJDHIE, both UEs are served in the unlicensed spectrum. Furthermore, in order to avoid interference to Wi-Fi users in the unlicensed spectrum, the transmit power of all users should satisfy

$$\sum_{i=1}^{M} \sum_{k=1}^{Q_i} h_{ikj} r_i \geq \max \left\{ \frac{B_u q_{j,s}}{p_{j,s}^{\max} q_{j,s} + 1}, \ \forall j \in \mathcal{N}, \forall s \in \mathcal{S} \right\}, \tag{34}$$

namely, in the example, the feasible region of the prices should be above EI and IH.

**Fig. 7** The feasible region of
the game

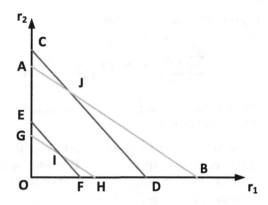

As all operators cooperate with each other, we assume that the prices set by all
operators satisfy

$$r_i = \theta_i r_1, \quad \forall i \in \{2, 3, \dots, M\}. \tag{35}$$

Substituting (35) into (25), we have

$$W_i = \sum_{s=1}^{S} \lambda_{j,s} \left( \theta_i \sum_{j=1}^{N} \frac{B_u \sum_{k=1}^{Q_i} h_{ikj}}{\sum_{l=1}^{M} \sum_{k=1}^{Q_i} h_{lkj}\theta_l} - r_i K_{i,s} \right), \tag{36}$$

where

$$K_{i,s} = \sum_{j=1}^{N} \frac{\sum_{k=1}^{Q_i} h_{ikj}}{q_{j,s}}. \tag{37}$$

Accordingly, the total utility of operators can be derived as

$$W^{all} = \sum_{i=1}^{M} \alpha_i \sum_{s=1}^{S} \lambda_{j,s} \left( \theta_i \sum_{j=1}^{N} \frac{B_u \sum_{k=1}^{Q_i} h_{ikj}}{\sum_{l=1}^{M} \sum_{k=1}^{Q_i} h_{lkj}\theta_l} - K_{i,s} r_i \right). \tag{38}$$

It is observed that when the relations of prices are fixed, the first part of $W^{all}$ in (38)
is not related to the value of prices. Based on the expression in the second part
of $W^{all}$, $W^{all}$ is linearly decreasing with each $r_i$, $\forall i \in \mathcal{M}$. Therefore, we have the
following lemma.

**Lemma 4** *The optimal solution to achieve the maximum $W^{all}$ lies in the boundary*

$$\sum_{i=1}^{M}\sum_{k=1}^{Q_i} h_{ikj}r_i \geq \max\left\{ \frac{B_u q_{j,s}}{p_{j,s}^{\max} q_{j,s} + 1}, \quad \forall j \in \mathcal{N}, \forall s \in \mathcal{S}\right\}. \tag{39}$$

*The position of the solution in the boundary depends on the parameters $K_{i,s}$, $\forall i \in \mathcal{M}$, $\forall s \in \mathcal{S}$ of prices.*

*Proof* When the UEs receive services in the unlicensed spectrum, in order to guarantee the performance of Wi-Fi users, the transmit power cannot be above the upper bound. Correspondingly, the price set by operators cannot be lower than the boundary

$$\sum_{i=1}^{M}\sum_{k=1}^{Q_i} h_{ikj}r_i \geq \max\left\{ \frac{B_u q_{j,s}}{p_{j,s}^{\max} q_{j,s} + 1}, \quad \forall j \in \mathcal{N}, \forall s \in \mathcal{S}\right\}. \tag{40}$$

Furthermore, when the prices of operators are coordinated, the total utility of operators is linearly decreasing with the price increasing. In order to achieve high utility of all operators, the prices of all operators decrease, and finally stop at the lowest boundary in (40). With different parameter $\theta_i$, the price decreases with different tracks, thus stoping at different positions in the lowest boundary.

We would like to find an optimal $K_{i,s}$, $\forall i \in \mathcal{M}$, $\forall s \in \mathcal{S}$ to achieve the maximum value of $W^{all}$, given the sub-band allocation results. We set the second part of $W^{all}$ as $G$, such as,

$$G = \sum_{i=1}^{M} \alpha_i K_{i,s} r_i. \tag{41}$$

Equation (41) is a hyperplane in the feasible region of prices. With $G$ increasing from a small value, the distance between the hyperplane and the feasible region decreases. Ultimately, the hyperplane will cut through the feasible region. The first point $O^*$ positioned $(r_1^*, r_2^*, \ldots, r_M^*)$ in the feasible region achieves the lowest value of $G$, compared with all other points in the feasible region. In other words, $O^*$ is the optimal point to achieve the maximum value of $W^{all}$. Correspondingly, the relationship of the prices follows

$$\theta_i = \frac{r_i^*}{r_1^*}. \tag{42}$$

To better understand this, we show the procedure in an example of two operators. Suppose there are two sub-bands in the unlicensed spectrum allocated to two UEs respectively. As shown in Fig. 8, the hyperplane is shown as $G = \alpha_1 K_{1,1} r_1 + \alpha_2 K_{2,2} r_2$. When $G$ approaches $G^*$, the hyperplane goes through the first point $O^*$

**Fig. 8** The optimal solution
when operators are
cooperative

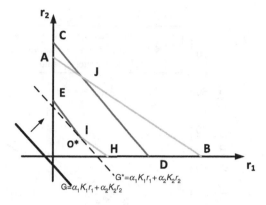

in the feasible region. As the position of point $O^*$ is $(r_1^*, r_2^*)$, $r_1^*$ and $r_2^*$ will be the optimal solution to achieve the maximum value of $W^{all}$. When the weight factors $\alpha_i$ in $W^{all}$ are different, the position of the optimal point $O^*$ may be different.

# 6  Simulation Results

We evaluate the performance of the proposed cooperative and non-cooperative scheme with MATLAB. We consider a hotspot circle area with a radius of 100 m. In the area, there are two operators, and each operator randomly deploys 2 SCBSs and 2 WAPs. We consider the uplink transmission and assume there are 100 UEs requesting service from the 20 sub-bands in the unlicensed spectrum. In order to avoid causing intolerably high interference to Wi-Fi users, we set the maximum transmit power of each UE in each time to be 2 W. We consider Additive White Gaussian Noise (AWGN) channels. Each sub-band in the unlicensed spectrum is 1 MHz, and the interference in each sub-band of the unlicensed spectrum for each UE is set as a random number with an average value of $-20$ dBm. The noise is assumed to be $-30$ dBm.

We first compare the performance of proposed cooperative and non-cooperative schemes with that of a single-operator scenario, where only one operator serves UEs in the unlicensed spectrum. As most existing resource management schemes in unlicensed spectrum assume a single-operator scenario, this comparison highlights the difference and advantages of our proposed strategies.

As shown in Fig. 9, we analyze the total utility of operators under different number of UEs. With an increasing number of UEs, the total utility of operators generally increases. In the proposed cooperative scheme, as the operators cooperate with each other, the total utility is the highest, followed by the non-cooperative scheme, where each operator makes decisions to maximize its own utility. Moreover, the total utilities in both the proposed cooperative and proposed non-cooperative schemes are higher than the total utility when there is only one operator in the scheme. In the single-operator cases, because of the limited number of WAPs, the total revenue received by the single operator is also limited.

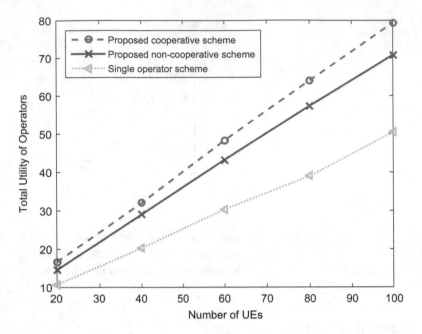

**Fig. 9** The total utility of operators vs. the number of UEs

In Fig. 10, the total utility of UEs under different numbers of UEs is studied. When the number of UEs increases, the total utility of UEs increases. In the proposed cooperated scheme, because of the cooperation of operators, the service prices set by the operators are low, and each UE can be served with high quality of service at low prices. Thus, the total utility of UEs is the highest. In the single operator scheme, the operator is able to set low price to all UEs, while each UE can choose the SCBSs from different base stations for better performance and lower prices. Therefore, the total utility of UEs with single operator scheme is higher than the utility in the proposed non-cooperative schemes, but lower than the utility in the proposed cooperative scheme. In the proposed non-cooperative scheme, due to the competition among operators, the prices set by operators do not reach the lower bound. Thus the UEs pay more to the operators, and the total utility of UEs keeps the lowest.

In Fig. 11, we analyze the total utility of operators under different number of WAPs of each operator. When the number of WAPs of both operators increases, for each WAP, each UE is required to pay the interference penalty. However, in the proposed cooperative scheme and single operator scheme, in order to avoid losing UEs because of the high interference penalty, the operators are able to reduce the price. Thus, with the number of WAPs increasing, the total utility of operators in the proposed cooperative scheme and single operator scheme generally does not change, while the total utility of operators in the proposed cooperative scheme keeps higher than the utility of operator in the single operator scheme. Moreover, in the

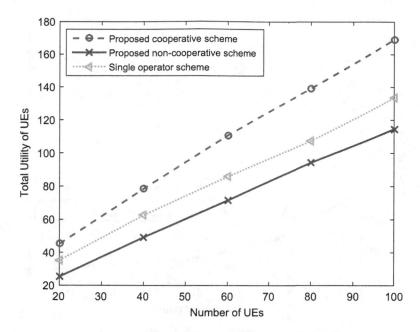

**Fig. 10** The total utility of UEs vs. the number of UEs

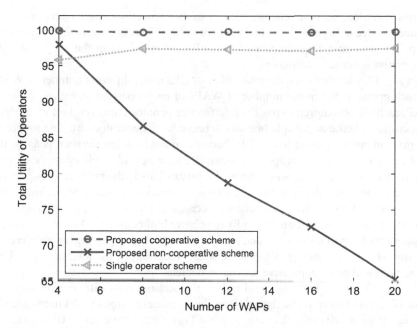

**Fig. 11** The total utility of operators vs. the number of WAPs of each operator

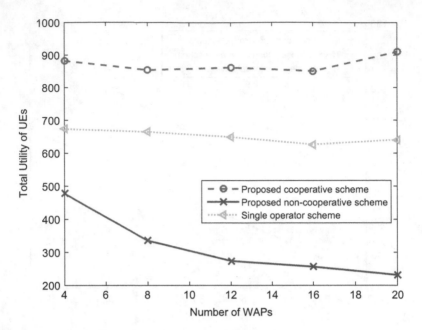

**Fig. 12** The total utility of UEs vs. the number of WAPs of each operator

proposed non-cooperative scheme, because of the competition, each operator cannot reduce its price unilaterally to achieve higher utility. Thus, the prices set by operators keep in high value. Therefore, the total utility of operators in the proposed non-cooperative scheme is decreasing.

In Fig. 12, we investigate the total utility of UEs under different number of WAPs of each operator. When the number of WAPs of each operator increases, for each WAP, each UE is required to pay the interference penalty. However, in the proposed cooperative scheme and single operator scheme, as the operators are able to reduce the price in order to avoid losing UEs because of the high interference penalty, the total utility of UEs in the proposed cooperative scheme and single operator scheme generally does not change, while the total utility of UEs in the proposed cooperative scheme keeps higher than the utility of UEs in the single operator scheme. Moreover, in the proposed non-cooperative scheme, because of the competition, each operator is unable to reduce its price unilaterally to achieve higher utility. Thus, the prices set by operators keep in high value, and each UE is supposed to pay higher interference penalty with the number of WAPs increasing. Accordingly, the total utility of UEs in the proposed non-cooperative scheme is decreasing.

In Fig. 13, we evaluate the total utility of operators with different interference from Wi-Fi. As shown in the figure, when the interference from Wi-Fi increases, the utilities of some UEs may decrease to zero. Therefore, with a fewer UEs using the unlicensed spectrum, the total utility of operators decreases. Accordingly, the total utility generally decreases. Moreover, for the proposed non-cooperative scheme, the

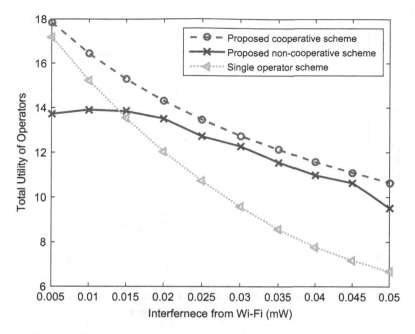

**Fig. 13** The total utility of operators vs. the interference from Wi-Fi

total utility of operators first increases slightly then decreases. The reason is that when the interference from Wi-Fi is small, the prices set by some operators may be very high. With the interference from Wi-Fi, the operators are able to reduce their prices first to motivate the UEs to purchase services in the unlicensed spectrum, and thus the utility increases. However, when the price reduces to the lowest boundary, in order to guarantee the performance of Wi-Fi users, the operators cannot reduce their prices anymore, and the utilities of UEs gradually reduce and reach zero ultimately. Moreover, the total utility of operators in the proposed cooperative scheme is always larger than the utility of the operators in the proposed non-cooperative scheme and the utility of the operator in the single operator scheme. When the interference from Wi-Fi is small, the prices set by the operators are high in the proposed non-cooperative scheme. Thus, the total utility of operators in the proposed non-cooperative scheme is lower than the utility of the operator in the single-operator schemes. With the interference from Wi-Fi increasing, the prices set by the operators in the proposed non-cooperative scheme gradually decreases. Thus, the total utility of operators in the proposed non-cooperative scheme gradually surpasses the utility of operator in the single-operator schemes.

In Fig. 14, we analyze the relation between the total utility of UEs with different interference from Wi-Fi. Because of the strong interference from Wi-Fi, some UEs may receive zero utility and refuse to be served in unlicensed spectrum. Accordingly, the utilities of UEs generally decrease. However, in the proposed non-cooperative scheme, because the operators can reduce their prices to motivate the

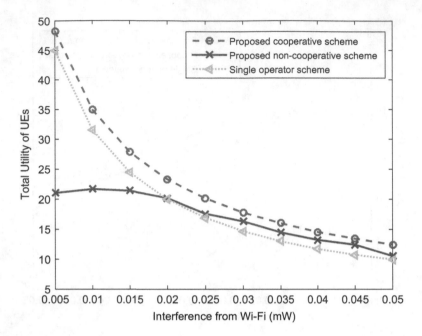

**Fig. 14** The total utility of UEs vs. the interference from Wi-Fi

UEs in the unlicensed spectrum, the utility of UEs first increases then decreases. The total utility of UEs in the proposed cooperative scheme is always larger than the utility of the UEs in the proposed non-cooperative scheme and the utility in the single operator scheme. When the interference from Wi-Fi is small, the prices set by the operators are high in the proposed non-cooperative scheme. Thus, the total utility of UEs is lower than the utility of UEs in the single-operator schemes. With the interference from Wi-Fi increasing, the prices set by the operators in the proposed non-cooperative scheme gradually decreases. Thus, the total utility of UEs in the proposed non-cooperative scheme gradually surpasses the utility of UEs in the single-operator schemes.

In Fig. 15, we discuss the relationship between the total utility of operators and the maximum transmit power of UEs. With the maximum transmit power increasing, as operators are able to serve UEs with a lower price, the total utility of operators generally increases. When the maximum transmit power of UEs are relatively small, In the proposed cooperative and non-cooperative scheme, as the UE is able to choose operators with higher quality of service and lower price, the total utility of operators in the proposed cooperative scheme and in the proposed noncooperative scheme are always larger than the utility in the single-operator scheme. Furthermore, because of the competition of operators, the prices set by the operators in the proposed cooperative scheme are relatively smaller than the prices in the proposed non-cooperative scheme. Thus, the total utility of operators in the proposed cooperative scheme remains higher than the utility in the

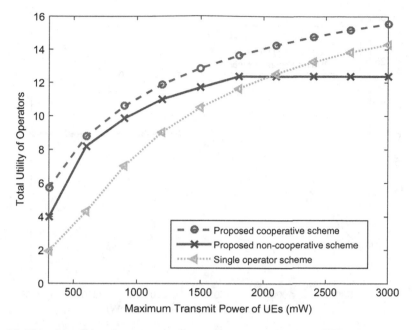

**Fig. 15** The total utilities of operators vs. the maximum transmit power of UEs

proposed non-cooperative scheme. Moreover, with the maximum transmit power increasing, the feasible region increases. When the Nash equilibrium point of the non-cooperative scheme is no longer in the boundary of the feasible regions, the total utility of operators in the proposed non-cooperative scheme stops increasing and keeps unchanged. Therefore, when the maximum transmit power is large, with the maximum transmit power increasing, the total utility of operators in the single operator scheme surpass the total utility of operators in the proposed non-cooperative scheme.

In Fig. 16, we analyze the relation between the total utility of UEs and the maximum transmit power of UEs. When the maximum transmit power increases, all UEs are able to transmit in high power, increasing the transmission rate during the service. Therefore, the total utility of all UEs generally increases. The total utility of UEs of the proposed cooperative scheme is always larger than that of the proposed non-cooperative scheme. Moreover, when the maximum transmit power is small, as the UE is able to choose operators with higher quality of service and lower price, the total utility of UEs in the proposed noncooperative scheme is larger than the utility in the single operator scheme. However, with the maximum transmit power increasing, the feasible region of in the Fig. 7 increases. When the Nash equilibrium point of the non-cooperative scheme is no longer in the boundary of the feasible regions, the total utility of UEs in the proposed non-cooperative scheme stops increasing and keeps unchanged. Therefore, when the maximum transmit power is large, with the

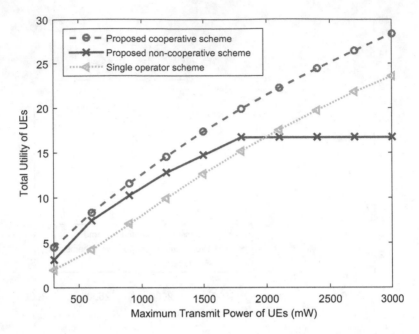

**Fig. 16** The total utilities of UEs vs. the maximum transmit power of UEs

maximum transmit power increasing, the total utility of UEs in the single operator scheme surpass the total utility of UEs in the proposed non-cooperative scheme.

In Fig. 17, we fix the value of $\alpha_2$ and increase $\alpha_1$ to evaluate the total utility of operators with different ratio $\alpha_1/\alpha_2$ of weight factors. In the simulated scenario, the ratios of the weight factor $\alpha_1/\alpha_2$ can be divided into five sections, which means that the first intersection $O^*$ of the hyperplane $G = \alpha_1 K_1 r_1 + \alpha_2 K_2 r_2$ and the feasible region fall in five different points based on different ratios of weight factor $\alpha_1/\alpha_2$. Within five sections, when the ratio increases, the total weighted utility of operators increases.

In Fig. 18, we evaluate the utility of operator 2 when its price decreases in both the proposed cooperative and non-cooperative schemes. As shown in the figure, in the proposed cooperative scheme, as the prices of operators are linearly related, with the price of operator 2 decreasing, the utility of operator 2 increases monotonically. Furthermore, in order to guarantee the basic data transmission of Wi-Fi users, when the prices of all other operators keep unchanged, there is a lower bound for the price set by operator 2. Therefore, the optimal price of operator 2 is the price in the lowest boundary. However, in the proposed non-cooperative scheme, when the price of operator 2 decreases and the price of operator 1 remains unchanged, the utility of operator 2 first increases and decreases. Thus, the optimal price of operator 2 is not in the lowest boundary in the non-cooperative scheme, but in the middle of the feasible region.

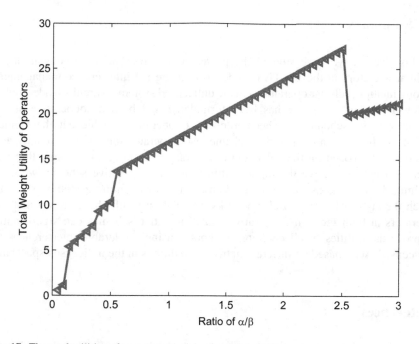

**Fig. 17**  The total utilities of operators vs. the ratio of weight factor

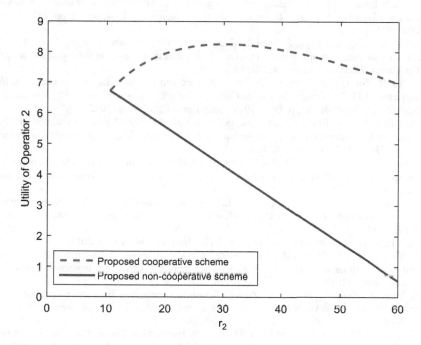

**Fig. 18**  The utility of operator 2 vs. price of operator 2

# 7  Summary

In this chapter, we have studied the power control mechanism among multiple cellular operators in the U-LTE in order to mitigate the interference management among multiple cellular operators and the unlicensed systems. A multi-leader multi-follower Stackelberg game has been formulated and both a cooperative and a non-cooperative schemes have been proposed for operators to achieve high revenues in U-LTE. In the non-cooperative scheme, each operator sets price rationally and independently based on the behaviors of others, and a sub-gradient algorithm has been adopted to achieve the highest utility. In the cooperative scheme, we have optimized the relations of the prices with a linear programming method so as to reach the highest utilities of all operators. Simulation results have shown that the operators in both the non-cooperative and cooperative schemes can significantly improve the utilities of all operators without causing intolerable interferences to unlicensed users, based on different network conditions in the unlicensed spectrum.

# References

1. Cisco, "Cisco Visual Networking Index: Global Mobile Data Traffic Forecast Update, 2016–2021 White Paper," ID:1454457600805266, Mar. 2017.
2. F. Liu, E. Bala, E. Erkip, M. Beluri, and R. Yang, "Small cell traffic balancing over licensed and unlicensed bands," *Vehicular Technology, IEEE Transactions,* vol. 64, no. 12, pp. 5850–5865, Dec. 2015.
3. J. Fu, X. Zhang, L. Cheng, Z. Shen, L. Chen, and D. Yang, "Guanding utility-based flexible resource allocation for integrated LTE-U and LTE wireless systems," *in Vehicular Technology Conference, IEEE 81st,* Glasgow, UK, May 2015.
4. Y. Xu, R. Yin, Q. Chen, and G. Yu, "Joint licensed and unlicensed spectrum allocation for unlicensed LTE," *Personal, Indoor, and Mobile Radio Communications, IEEE 26th Annual International Symposium,* pp. 1912–1917, Hong Kong, China, Aug. 2015.
5. O. Sallent, J. Perez-Romero, R. Ferrus, and R. Agusti, "Learning-based coexistence for LTE operation in unlicensed bands," *Communication Workshop, IEEE International Conference on,* pp. 2307–2313, London, UK, Jun. 2015.
6. Y. Gu, Y. Zhang, L. X. Cai, M. Pan, L. Song, and Z. Han, "Student-project allocation matching for spectrum sharing in LTE-Unlicensed," *Global Communications Conference, IEEE,* San Diego, CA, Dec. 2015.
7. H. Zhang, Y. Xiao, L. X. Cai, D. Niyato, L. Song and Z. Han, "A hierarchical game approach for multi-operator spectrum sharing in LTE unlicensed," *Global Communications Conference, IEEE,* San Diego, CA, Dec. 2015.
8. "U-LTE: unlicensed spectrum utilization of LTE," *Huawei White Paper,* 2014.
9. Y. Xiao, Z. Han, C. Yuen and L. A. DaSilva, "Carrier aggregation between operators in next generation cellular networks: A stable roommate market," *Wireless Communications, IEEE Transactions,* vol. 15, no. 1, pp. 633–650, Jan. 2016
10. P. Yuan, Y. Xiao, G. Bi and L. Zhang, "Towards cooperation by carrier aggregation in heterogeneous networks: A hierarchical game approach," in *IEEE Transactions on Vehicular Technology,* vol. 66, no. 2, pp. 1670–1683, Feb. 2017.
11. Y. Xiao, G. Bi, D. Niyato and L. A. Dasilva, "A Hierarchical Game Theoretic Framework for Cognitive Radio Networks," *IEEE Journal on Selected Areas in Communications,* vol. 30, no. 10, pp. 2053–2069, Nov. 2012.

12. Y. Xiao, D. Niyato, Z. Han and K.-C. Chen, "Secondary users entering the pool: A joint optimization framework for spectrum pooling," *IEEE Journal Selected Areas in Communications*,
vol. 32, no. 3, pp. 572–588, Mar. 2014.
13. Y. Shen and E. Martinez, "Channel estimation in OFDM systems," *Application note, Freescale
semiconductor,* 2006.
14. D. Gale and L. S. Shapley, "College admissions and the stability of marriage," *American
Mathematical Monthly,* vol. 69, no. 1, pp. 9–15, Jan. 1962.
15. D. G. McVitie and L. B. Wilson. "The stable marriage problem," *Communications of the ACM,*
vol. 14, no. 7, pp. 486–490, Jul. 1971.
16. S. P. Boyd and L. Vandenberghe, *"Convex optimization," Cambridge University Press,* 2004.
17. Y. Xiao, G. Bi, and D. Niyato, "A simple distributed power control algorithm for cognitive radio
networks," *Wireless Communications, IEEE Transactions,* vol. 10, no. 11, pp. 3594–3600, Nov.
2011.

# Conclusions and Future Works

## 1 Conclusions and Remarks

U-LTE provides users with high quality of service, yet it may generate strong interference to other unlicensed systems if no proper resource allocation is performed. In this book, we thoroughly investigate and improve on the coexistence technologies of U-LTE with other unlicensed systems over the unlicensed band. Specifically, we first analyze the performance of existing coexistence technologies, and identify key issues in terms of fairness and protocol overhead. A new coexistence mechanism has been carefully designed to achieve a more efficient and harmonious coexistence among U-LTE and Wi-Fi over the unlicensed band. Further analysis is performed to optimize the performance of the proposed coexistence technology, which has also been validated by system-level simulations.

Next, spectrum sharing in the unlicensed spectrum is analyzed in the multi-operator scenario. Considering the distributive behaviors of all wireless service operators and other unlicensed systems, game theory is employed to analyze the behaviors of each individual. In order to guarantee the performance of original unlicensed systems, each wireless service operator first adopts a zero-determinant strategy during the interaction with other unlicensed systems. As the behaviors of each wireless service operator can influence the utilities of others based on the restricted behaviors from the zero-determinant strategy, a non-cooperative game is formulated, where each wireless service operator determines its optimal strategy and achieves the Nash equilibrium result.

During the spectrum sharing, due to multiple unlicensed bands available the U-LTE and other unlicensed systems, the stable marriage (SM) game in matching theory is proposed for the interaction between LTE and Wi-Fi users, and the coexistence constraints are interpreted as the preference lists. In the SM game, two semi-distributed solutions, namely the Gale-Shapley (GS) and the Random Path to Stability (RPTS) algorithms are proposed. In order to address the external effect

© The Author(s) 2018
H. Zhang et al., *Resource Allocation in Unlicensed Long Term Evolution HetNets*,
SpringerBriefs in Electrical and Computer Engineering,
https://doi.org/10.1007/978-3-319-68312-6_6

in matching, the Inter-Channel Cooperation algorithm is introduced. The resource allocation problem is studied with network dynamics, and the proposed mechanisms are evaluated under two typical user mobility models.

Furthermore, resource allocation of both the licensed and the unlicensed spectrum is analyzed from the perspective of each wireless service operator. In the unlicensed spectrum, the behaviors of one wireless service operator can affect the utilities of other wireless service operators and users. Accordingly, a multi-operator multi-user Stackelberg game is proposed, where all wireless service operators act as leaders and all users act as followers. In order to avoid intolerable interference to the Wi-Fi access point (WAP), each operator sets an interference penalty price for each user that causes interference to the WAP. Based on which, users can choose their sub-bands and determine the optimal transmit power in the chosen sub-bands of the unlicensed spectrum. With the first-mover advantage, wireless service operators can predict the behaviors of each user. Considering the behaviors of other wireless service operators with both coordinative and competitive relations, the optimal pricing strategy for each wireless service operator is proposed. Simulation results are further presented to demonstrate the improved performance of our proposed schemes.

## 2 Future Works

### 2.1 Multi-Channel Aggregation in U-LTE

The existing U-LTE MAC protocols only consider spectrum sharing of single cell U-LTE and Wi-Fi over a single unlicensed channel. However, the recent IEEE 802.11ac and 802.11ax Wi-Fi standards, specified an enhanced MAC protocol that allows a user to opportunistically bond multiple channels for high data rate transmissions. Unlicensed LTE is inherently a multi-carrier network, where multiple unlicensed channels can be aggregated for resource allocation at the LTE BSs. To capture this feature, we need to design a multi-channel MAC protocol for U-LTE, and find the best channel access strategies for both U-LTE and Wi-Fi, to attain the best network performance while ensuring fair sharing of the unlicensed users.

### 2.2 Explorations of Unlicensed Spectrum in mmWave

Besides the 5 GHz frequency band, there is a growing interest to integrate unlicensed mmWave spectrum bands into LTE, in order to exploit the ultra-wide bandwidth available at the unlicensed 60 GHz band and the E-band (71–76 and 81–86 GHz). Advanced mmWave communication has many salient features and provides a promising solution for high density multi-band HetNets, yet it also faces new challenges when integrated into the existing cellular infrastructure, such as ensuring inter-operability with existing LTE services. How to exploit mmWave communications in the design and implementation of U-LTE heterogeneous networks remains an open research issue. For example, directional transmissions in mmWave band usually require Line of Sight (LOS) communication links, as mmWave signals are very sensitive to blockage effects due to the severe penetration loss. Thus, in a mmWave based U-LTE where the LOS link may be blocked from time to time, it is desirable to allow one user to associate with multiple BSs to ensure service continuity. Furthermore, it is also promising to implement relay networks for data transmission in the blocked areas. Accordingly, how to motivate BSs or relays to receive data service, and how to design algorithms to select an appropriate set of BSs or relays for serving users at a reasonable complexity, are interesting yet challenging research issues.

## 2.3  Network Virtualization for Resource Allocation in U-LTE

With increasing variety of data services, the variety of service operators increases correspondingly. As different service providers have different purpose and requirements when exploiting data service in unlicensed spectrum, one or multiple spectrum providers are required to adopt the mapping between different service providers and all kinds of unlicensed spectrums, so as to make flexible resource allocation and improve the network efficiency. Due to the complexity of the mapping, the service operators are invisible to the resource management in the unlicensed spectrum, and just apply the appropriate amount of unlicensed spectrum with low interference for its data services in the virtualized networks. Nevertheless, due to the spectrum reuse and spectrum aggregation issues, considering the different preferences of service operators and unlicensed spectrum, the management of resource allocation is required. Moreover, when there are three layers consisting of multiple service providers, multiple spectrum providers and multiple unlicensed spectrum bands, respectively, how to deal the competition within each layer and mapping between each two layers in distributive fashion remains challenging.

Printed in the United States
By Bookmasters